Springer Theses

Recognizing Outstanding Ph.D. Research

Aims and Scope

The series "Springer Theses" brings together a selection of the very best Ph.D. theses from around the world and across the physical sciences. Nominated and endorsed by two recognized specialists, each published volume has been selected for its scientific excellence and the high impact of its contents for the pertinent field of research. For greater accessibility to non-specialists, the published versions include an extended introduction, as well as a foreword by the student's supervisor explaining the special relevance of the work for the field. As a whole, the series will provide a valuable resource both for newcomers to the research fields described, and for other scientists seeking detailed background information on special questions. Finally, it provides an accredited documentation of the valuable contributions made by today's younger generation of scientists.

Theses are accepted into the series by invited nomination only and must fulfill all of the following criteria

- They must be written in good English.
- The topic should fall within the confines of Chemistry, Physics, Earth Sciences, Engineering and related interdisciplinary fields such as Materials, Nanoscience, Chemical Engineering, Complex Systems and Biophysics.
- The work reported in the thesis must represent a significant scientific advance.
- If the thesis includes previously published material, permission to reproduce this must be gained from the respective copyright holder.
- They must have been examined and passed during the 12 months prior to nomination.
- Each thesis should include a foreword by the supervisor outlining the significance of its content.
- The theses should have a clearly defined structure including an introduction accessible to scientists not expert in that particular field.

More information about this series at http://www.springer.com/series/8790

José Juan Colás

Dual-Mode Electro-photonic Silicon Biosensors

Doctoral Thesis accepted by
University of York, UK

Springer

Author
Dr. José Juan Colás
Department of Physics
University of York
York
UK

Supervisor
Dr. Steven D. Johnson
University of York
York
UK

ISSN 2190-5053 ISSN 2190-5061 (electronic)
Springer Theses
ISBN 978-3-319-60500-5 ISBN 978-3-319-60501-2 (eBook)
DOI 10.1007/978-3-319-60501-2

Library of Congress Control Number: 2017944313

Printed on acid-free paper

This Springer imprint is published by Springer Nature
The registered company is Springer International Publishing AG
The registered company address is: Gewerbestrasse 11, 6330 Cham, Switzerland

To my parents

Supervisor's Foreword

The enzyme-linked immunosorbent assay has been the cornerstone of in vitro protein diagnostics for over 40 years and is used throughout health care for the detection and quantification of disease biomarkers in clinical samples. The demand for such tests is, however, increasing rapidly and cannot be met by current technologies in the near future. The driving force behind the increased demand is in part the increasing number and age of the population. Recent advances in medical science and new approaches to the management of diseases are also redefining what is required from clinical testing. For example, a new generation of drugs, known as targeted therapies, are increasingly being used in cancer treatment. These drugs show remarkable therapeutic effect without the adverse side effects common to many conventional medicines. However, these drugs are so highly specific to a certain type of a disease that they no longer work for the majority of patients, even though symptoms may be similar. There is therefore a growing need for new tests that can inform clinicians about the most appropriate treatment choice for an individual patient. But, developing these tests is becoming progressively complex as the next generation of therapies become increasingly targeted. Hence, many different tests have to be carried out simultaneously to obtain accurate information on the patients disease.

Label-free, silicon photonic biosensors are widely considered to be a potential technological solution to these challenges. These analytical devices consist of a highly specific, high-affinity probe molecule, most commonly an antibody, that is interfaced with a physiochemical transducer that is sensitive to the local refractive index. A change in refractive index, for example, due to formation of the antibody–antigen complex, can thus be monitored directly, i.e., without the need for an additional antibody, enzymatic, fluorescent or radioactive label, to provide a response that is proportional to protein concentration.

The multiplexed detection of tens or even hundreds of thousands of protein species in a small clinical sample is a significant technological challenge that requires the identification of new approaches for fabricating high-density, submicrometer-scale photonic biosensor arrays. Many existing silicon photonic biosensors are fabricated using the same low-cost, mass-manufacturing techniques developed in the semiconductor industry. These approaches readily provide the submicrometer-scale resolution required for the fabrication of high-density biosensor arrays; however, the technology required to functionalise each biosensor within the array with a different probe molecule is very much in its infancy. For example, dot-printing approaches for creating antibody arrays are currently spatially limited to around 0.1 mm. Although the resolution of these printing techniques is likely to improve, the feature sizes required for high-density biosensor arrays capable of handling extremely small sample volumes are likely to remain beyond the scope of such systems. While high spatial resolution molecular immobilisation has been demonstrated, for example, using nanografting or dip-pen nanolithography, these serial approaches do not provide the high-throughput required for mass fabrication of high-density protein arrays.

Electrochemically controlled functionalisation is a promising approach to spatially direct the immobilisation of biomolecules that simultaneously meets the requirements of resolution, speed and the ability to coat each biosensor within an array with a different probe molecule. While this has been demonstrated for electrochemical biosensors, there exists an apparent challenge in implementing this in silicon photonic sensors; high sensitivity photonic detection of proteins demands a silicon substrate with low doping density, while silicon electrochemistry requires high doping density. This thesis discusses a practical solution to this challenge and demonstrates an electro-photonic biosensor capable of supporting both photonic and electrochemical measurements in a single technology.

José Juan-Colás was key to this innovation. José came to York having conducted his Master's project in the well-known Ghent Photonics group, who pioneered the foundry approach for silicon photonic devices. Making devices in a foundry instead of a local cleanroom is well established in the microelectronics industry through organisations such as Europractice, who offer shared access to high-end technology for a modest cost. Experiencing the foundry model and working at Ghent gave Pepe a solid grounding in silicon photonic devices, especially microring resonators. It highlighted the importance of careful device design to him, as the turnaround time even in Gent is several months, so a student, especially a Master's student, cannot afford to make a mistake as they may not have a second chance to correct it.

However, José wanted to spread his wings and to explore silicon photonics further. He was particularly attracted by the proposition to conduct exploratory research and to collaborate with researchers in different disciplines, especially Chemistry and Biology. While this interdisciplinary approach to research is widely recognised as valuable and important, particularly in biosensor development, achieving it in practice is extremely challenging. However, through a combination of extreme hard work, academic learning and persistence, José demonstrated remarkable success, as is well documented in the following pages.

York, UK
June 2017

Dr. Steven D. Johnson

Abstract

Our increased understanding of the molecular biology of disease has had a significant impact on health care. Genetics has allowed the identification of hereditary diseases and predisposition for others, such as cancer. However, for many diseases, it is also necessary to monitor the expression of panels of proteins. The need to monitor protein expression presents a significant technological challenge requiring a highly multiplexed analytical technology with a sensitivity down to femtomolar range.

Low-cost photonic devices are highly sensitive to changes in their local environment and can be chemically modified to exhibit high specificity detection towards, for instance, proteins or DNA oligonucleotides. However, even if small-footprint photonic biosensors can be engineered in silicon microarrays, approaches to realise the very high-density, multiplexed sensing potential of photonic biosensors have yet to be demonstrated. This study aims to develop and demonstrate, for the first time, a dual-mode electro-photonic technology capable of highly multiplexed detection at the submicron scale and multiparameter profiling of biomolecules on the silicon photonics platform. Furthermore, the technology integrates electrochemical and photonic measurements in a single sensor platform. By combining the complementary information revealed by each of the domains, it is possible to broaden the range of systems that are accessible for silicon photonics.

Our dual-mode technology consists of microring resonators optimally n-doped (doping density of 7.5×10^{16} cm^{-3}) to support high-Q resonances (Q-factor $\approx$ 50,000) alongside electrochemical processes in situ. This combination of sensing mechanisms enables the application of electrochemical methods for site-controlled immobilisation of receptor molecules. Furthermore, electrochemical characterisation of molecules bound to the sensor surface also provides direct quantification of binding density and unique insight into chemical reactivity, which is unavailable with photonic detection alone. This unique technology, based on the combination of electrochemical and photonic sensing on a silicon platform, not only enables detection of multiple biological molecules required for future clinical diagnostics, but also has the potential to impact on fundamental biochemical research.

Acknowledgements

My sincerest gratitude goes out to all those who have contributed to the success of this interdisciplinary work. This would not have been possible without their constant and precious support.

First of all, I would like to thank my supervisors Dr. Steve Johnson and Prof. Thomas Krauss, who have greatly influenced both my professional and personal development. Beyond supervision, I could not have wished for better mentoring and encouragement throughout my doctorate. Knowing that I could always discuss ideas helped me to look forward and think of 'What's next?'. Steve, thanks for your lively, positive attitude and never ending support throughout these three years. Thomas, thanks for your friendly and close guidance, and providing me with precious input, I am truly grateful for it. The acknowledgement extends to Dr. Alison Parkin for her support and advice on the chemistry component of my research.

I would like to thank all my colleagues and friends from the Photonics group and those from the Bio-Inspired Technologies group. My appreciation goes to Ahmed Ben Khaial, Donato Conteduca, Dr. Annett Fischer, Dr. Matthias Fischer, Elena Koutsoumpeli, Dr. Bryan O'Regan, Giampaolo Pitruzzello, Dr. Daan Stellinga, Dr. Mark Scullion, Dr. Christopher Reardon, Lewis Reeves, Dr. Yue Wang and Dr. Davide Zecca, who have all contributed to a warm environment and represented my 'York family' during my Ph.D. I am glad I have shared this experience with you all. Special thanks goes to Dr. Katherine Dunn for her priceless practical advice, particularly in the last year.

Finally, I would like to thank all those fabulous friends I have met during the course of the last three years and make York my second home for these years. Per descomptat, no oblide a la gent que des de casa ha estat donant-me suport a tot moment. Els meus pares, germanes, cunyats i 'tio Paco', per marcar-me el camí i per haver sabut transmetrem eixa energia que tantes vegades he necessitat. Als meus bons amics, que tant m'han animat als bons i roïns moments. I per últim, i no per això menys important, a Paula per compartir tots estos anys, els matins d'esport i les nits de soparets, les pujades i les baixades, els dies fàcils i els no tan fàcils… i per damunt de tot, per fer-ho possible.

Contents

Chapter 1
Preamble

In this chapter I introduce the study carried out in this thesis. It explains the motivation that originated this work, followed by a description of the objectives and the outline of this manuscript.

1.1 The Need for Biosensors

A single droplet of blood, which may have a diameter of 2 mm (i.e. a volume of 4 microliters), contains approximately 20 million cells, 20 thousand of which contain the genetic code for the creation of the 30 thousand proteins of the human genome [1]. A large number of these proteins will be present in the droplet, inside the blood cells and on their membrane, but also suspended in the fluid. A number of these proteins, known as biomarkers, have been identified as useful indicators of disease. Many patient decisions today rely on our ability to detect such biomarkers, or more accurately to quantitatively measure of the absolute biomarker concentration. The key component in biomarker detection is the biosensor, which translates the recognition of biomarkers into a quantifiable signal that can be fed back to the clinician to provide a better insight into a patient's disease and inform appropriate treatment options. Despite considerable progress in understanding the genetic basis of disease, genetic information alone is insufficient to fully understand or diagnose disease. The genetic code contains information about protein synthesis however in some cases abnormalities occur during or after protein translation. The study of this dynamic protein behaviour is commonly referred to as proteomics [2] and is often more complex than genomics (the study of genome). Firstly, a technique to study proteomics should be highly multiplexed, which means that a large number of molecular interactions must be monitored simultaneously, as frequently, a disease alters the expression of multiple proteins. Secondly, the technique should be capable of detecting proteins over a wide range of concentrations down to femtomolar concentrations, while the sample

J. Juan Colaś , *Dual-Mode Electro-photonic Silicon Biosensors*, Springer Theses,
DOI 10.1007/978-3-319-60501-2_1

contains high levels of salt and molecules at concentrations up to several hundreds of micromolars. Thirdly, different proteins have very different properties, further complicating the analysis. Additional requirements for the underpinning technology are high throughput, such that a large number of samples can be analysed in a short time frame, quantitative measurements, simple sample preparation and low-cost.
Many different types of sensor exist, most of them optimised for one particular role, and with different transduction mechanisims (i.e. sensing modes). Some of the most common transduction mechanisms rely on a change in electrical, optical, chemical, acoustic and mechanical, or thermal properties. In some cases there can even be combinations of these different mechanisms. Many sensors in the past have relied on electrical transduction mechanisms benefiting from the technological developments of the electronics manufacturing industry, and the correspondingly cheap components and fabrication techniques available. More recently, advances in optics and photonics technology have led the introduction of sensors based on optical transduction. Indeed, sensors based on optical transduction mechanisms offer a high degree of multiplexing and throughput, and have been proven to provide accurate, fast and reliable analysis (see Chap. 2). In contrast to their electrical counterparts, they offer certain advantages: they are immune to electromagnetic interferences, they can be designed so that they have well defined spectral features that can be engineered in a device as a point of reference, and also many media have unique optical absorption and scattering spectra that can act as a 'fingerprint to help identify them. Furthermore, they can be designed in a way that physical contact between different parts of the sensor is not needed, opening the prospect of further integration of filters, electronics, sources and detectors. Such capabilities underpin the potential for portable instruments, capable of high-end point-of-care diagnostics. However, approaches to realise the very high-density, multiplexed sensing potential of optical sensors have yet to be demonstrated. Here we show an innovative sensor technology that integrates electrical with optical technology and that is capable of combining the two sensing modes in multiplexed format, not only providing deeper insight into the undergoing molecular and (bio)molecular processes, but also allowing the fabrication of very high-density photonic sensor arrays.

1.2 Dual-Mode Silicon Electro-Photonic Biosensors

Biophysical tools that combine multiple transduction mechanisms are becoming increasingly popular within the scientific community, as they can simultaneously probe different properties of the (bio)molecules under study and broaden the range of systems that are accessible for analysis. For instance, technology that combines acoustic and electrochemical sensing (Sect. 2.4.2) provide real-time information on the mass, viscoelastic and electrochemical properties of the (bio)molecules under study. Similarly, combinations of optical and electrochemical sensing provide insight into both the structure and electrochemical properties of electroactive molecules, beyond what is possible with optical detection only.

The development of nanofabrication technology in the past decades has allowed the fabrication of tailored structures on the nanometer scale, some of which has been employed in the fabrication of biosensors. Specifically, photonic nanostructures constructed on a similar scale as the wavelength of light (nanometer scale), and in a material with a high refractive index such as silicon ($n_{silicon} \approx 3.44$), can provide a very attractive platform for sensors. In this platform, commonly known as silicon photonics, customised nanostructures with submicrometer dimensions are extremely sensitive to variations in their environment. For example, resonant structures can be fabricated which show sharp resonance peaks in the transmission spectrum [3, 4]. The position of these peaks will shift in wavelength due to local changes to the environment. For instance, these waveguides can be coated with immobilised receptor molecules that bind their complementary molecules (e.g. biomarkers) onto the surface when a given fluid (such as serum or blood) is delivered to this surface. This binding event will change the effective refractive index on the waveguide surface (due to the overlap of the evanescent wave that penetrates into the supporting solution) transferring this change to the waveguide's optical mode. In this way, the silicon nanostructure acts as a transducer, converting a binding interaction into an optical signal.

Incorporating the modality of electrochemical interrogation into the silicon photonics platform would provide additional functionality that until now been inaccessible to silicon photonics. For example, this integration would enable electrochemical reactions to be controlled and monitored on the silicon surface. This not only provides insight into electrochemical processes but can also be exploited for chemical modification of the photonic sensors. In this manner, this capability can be exploited to direct the immobilisation of different receptor molecules onto individual sensors within a silicon photonic array, underpinning the fabrication of very high-density photonic sensor arrays. Employing the surface modification process reported in this work, we believe a resolution of 15 nm can be achieved [5]. No other technique for introducing probe molecules locally onto a silicon photonic surface meets simultaneously the requirements of resolution, speed, and the ability to coat different sensors with a specific binding molecule.

Amongst different photonic resonant structures, biosensors based on microring resonators have seen much success in both research and commercial applications due to the fact that they are tolerant to fabrication deviations and exhibit very sharp resonances (i.e. high resonance quality factors) (Sect. 2.3.2.4). In general, ring resonators consist of ring-waveguide and a coupling mechanism to access the ring loop. When the ring waveguide accommodates an integer number of the wavelength, coupling to the ring becomes very strong and a sharp dip in transmission occurs. When the ring cavity is coated with immobilised receptor molecules, subsequent binding of the complementary biomolecules causes a small refractive index change that shifts the resonance wavelength. In this work, our electro-photonic technology consists of high quality factor (Q-factor $\approx$ 50,000), optical ring resonators fabricated on silicon-on-insulator wafers, which have been optimally n-doped (doping density of $7.5 \times 10^{16} cm^{-3}$) to support electrochemical processes (Fig. 1.1). The sensitivity

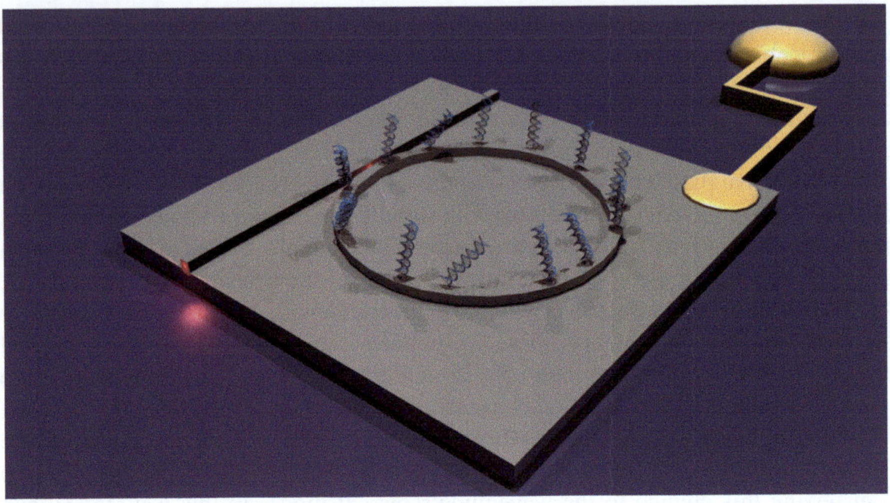

Fig. 1.1 Silicon electro-photonic ring resonator. The silicon layer has been optimally n-doped to support electrochemical processes (Chap. 4)

of the device, measured to be 65.1 nm/RIU, leads to a Limit Of Detection (*LOD*) of 1.15×10^{-5} RIU, comparable to ring resonator sensors fabricated in industrial foundries and published in the literature.

1.3 Goals of This Thesis

Given the potential advantages of combining two different sensing modes, this thesis describes the realisation of a dual-mode electro-photonic biosensor on a silicon photonics platform. In addition, the thesis aims to highlight the benefits of this approach over other similar bi-domain techniques by demonstrating its capabilities for solving real molecular and biological problems. The interdisciplinary nature of this work means that much effort was taken to collaborate with experts from across the sciences and other potential end-users of the devices fabricated on this platform. A number of fundamental challenges were addressed in the realisation of this dual-mode biosensor:

Incorporation of electrochemical interrogation into the silicon photonics platform

Since the requirements for electrical sensing are different to those for optical sensing, approaches for efficiently adding electrochemical control to a photonic sensor were investigated. A suitable approach which does not limit the performance of the photonic sensor and that is compatible with standard CMOS fabrication processes is therefore desired.

Ability to functionalise with a wide range of chemical moieties

In order to specifically detect a biomarker of interest, suitable receptors have to be identified which can be chemically attached to the silicon sensor surface. Due to the wide spectrum of biomolecules of interest for this research, compatibility with a range of different chemical families is therefore required.

Integration of a microfluidics platform

Microfluidic channels are required to ensure the delivery of samples to the sensor surface, minimise sample volumes and prevent evaporation. An approach for incorporating microfluidics into the final device is needed.

Specific detection of biomolecules

A final device must be able to detect the target biomolecules in real-time with high sensitivity. In addition, specificity is required, as in real applications, the target biomolecules will not be isolated in the sample.

Small footprint

Ideally, if a highly multiplexed sensor is to be fabricated for point-of-care use, the sensors should have small dimensions to allow further integration with devices such as sources and detectors to provide a fully working device.

Provide insight into biological problems of interest

Another goal is to demonstrate that our platform provides a novel method for exploring complex biological scenarios and further insight than has been hitherto possible.

1.4 Outline of the Thesis

This thesis is structured into seven chapters. A brief synopsis of these chapters is as follows:

Chapter 1 Introduction

In this chapter the thesis is introduced, highlighting its motivation, goals and the outline of this manuscript.

Chapter 2 Fundamentals of Label-Free Biosensing

This chapter discusses the fundamental concepts related to label-free biosensors and the physics behind the underpinning optical sensing mechanism. Alternative dual-mode techniques are also introduced together with details about their operational principle.

Chapter 3 Fabrication and Experimental Techniques

The fundamentals of the nanofabrication and experimental techniques employed throughout this thesis are presented in this section.

Chapter 4 The Electro-Photonic Silicon Biosensor

Our dual-mode sensor is presented. An in-depth characterisation of its properties is given alongside a proof-of-concept experiment. The device is compared to similar bi-domain sensing techniques.

Chapter 5 Study and Application of Electrografted Layers of Diazonium Ions

The inclusion of electrochemical interrogation is introduced and used to exploit electrochemical processes, which are ultimately employed to selectively modify the silicon surface and to regulate the local surface chemistry.

Chapter 6 Tailoring Light-Matter Interaction for Quantification of Biological and Molecular Layers

A further capability of our platform presented is discussed, which provides further, quantitative information regarding the structure (refractive index and thickness) of surface immobilised molecular layers. This is demonstrated for the monitoring of conformational changes of bound molecules.

Chapter 7 Conclusions and Outlook

This last section collects all of the conclusions of this work and discusses future prospects of the technique.

References

1. J. Lewis, M. Raff, K. Roberts, B. Alberts, A. Johnson and P. Walter. Molecular biology of the cell. Garland Sci. (2002)
2. S. Hanash, Disease proteomics. Nature **422**, 226–232 (2003)
3. R. Puchades, M.-J. Bauls, N. Maquieira, Chemical surface modifications for the development of silicon-based label-free integrated optical (io) biosensors: a review. Analytica Chimica Acta **777**, 1–16 (2013)
4. X. Fan, Y. Sun, Optical ring resonators for biochemical and chemical sensing. Anal. Bioanal. Chem. **399**(1), 205–211 (2011)
5. S. Palacin, J. Leroy, F.D. Del, L. Zagonel, N. Barrett, O. Renault, A. Bailly, J. Charlier, D. Mariolle, Local silicon doping as a promoter of patterned electrografting of diazonium for directed surface functionalization. J. Mater. Chem. **18**, 3136–3142 (2008)

Chapter 2
Introduction to Label-Free Biosensing

This section of the thesis introduces the fundamental concepts related to label-free biosensing and the theoretical concepts related to the resonant photonics employed in this thesis. Starting with an introduction to conventional approaches of molecular detection and medical diagnostics, a range of single-mode label-free biosensing techniques are then presented. The theoretical frameworks that underpin photonic waveguide-based detection strategies are discussed, which then serves to understand and explain the sensing principle of the devices discussed in later chapters. This is followed by an introduction to the most widespread dual-mode sensing principles, namely electrochemical detection combined with optical or acoustic sensing. A presentation of dual-mode electro-photonic silicon biosensing is then given before concluding the chapter with an introduction to functionalisation techniques using self-assembled monolayers.

2.1 Motivation

Biosensors that are able to detect disease specific molecular biomarkers play a crucial role across healthcare, from initial diagnosis to optimising and monitoring treatment. Such biosensors need to be able to quantify biological molecules specifically and selectively and with high temporal resolution to accurately monitor their evolution. This combination of requirements can be extremely challenging as a large number of different molecules are present in a clinical sample while only a single one, or a small group of molecules being markers of a specific disease. In addition, it is envisaged that with the introduction of personalised medicine, biomarker profiles rather than individual biomarkers will be required. A highly sensitive, selective and multiplexed biosensing platform is therefore needed to meet all of these requirements [1].

Moreover, many biochemical analytes, including DNA, RNA, proteins, viral capsides, and small molecules are sometimes present at concentration orders of fg/ml

J. Juan Colaś, *Dual-Mode Electro-photonic Silicon Biosensors*, Springer Theses,
DOI 10.1007/978-3-319-60501-2_2

to pg/ml together with other molecules. Biosensor technologies must therefore be highly sensitive.

Detecting biological analytes directly and specifically according to physical properties (such as physical size, mass or charge) is extremely challenging. Most biochemical assays thus exploit the high affinity of a 'receptor molecule' towards a specific biomarker. For example, in the case of protein detection, this receptor molecule is typically an antibody raised against the specific protein antigen while a complementary single stranded DNA is used for specific DNA detection. Following binding to the target molecule, the receptor molecule can be functionalised with a 'label' that is easy to measure and quantify, for example an enzyme that produces a colorimetric response as in the well established *Enzyme-Linked ImmunoSorbent Assay* (ELISA) [2] technology which enables the detection and quantification of specific antigens in a sample (Fig. 2.1). Particularly, detection at subfemtomolar concentrations has been reported using this assay [3].

The first step in an ELISA assay (Fig. 2.1, 1) is the immobilisation of antibodies at the surface of a well (typically these kind of assays are performed in microwell plates or reaction tubes). The sample containing the complementary antigen is added, leading to the formation of antibody-antigen complexes (Fig. 2.1, 2). Thereafter, the surface is washed with a detergent solution that removes material bound non-specifically to the surface. After this, a second monoclonal antibody which binds to a different region of the target is added (Fig. 2.1, 3). This secondary antibody is modified to carry a reporter enzyme designed to produce a color change when the enzyme reacts with its substrate. If the antigen is present, a complex will have been formed that includes the antibody bound to the well, the antigen and the enzyme-conjugated antibody (Fig. 2.1, 4). To conclude the assay, the specific enzyme substrate is added to produce a visible signal proportional to the quantity of antigen present in the sample (Fig. 2.1, 5).

Even though labels have been essential for implementing nearly all biochemical and cell based assays, this technique presents several practical drawbacks. Firstly,

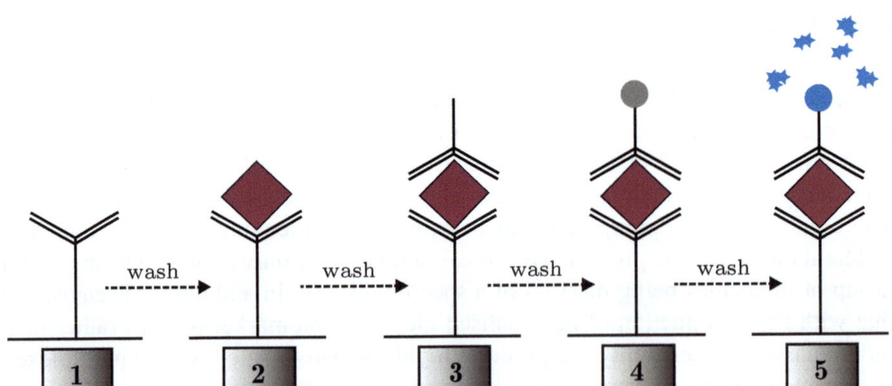

Fig. 2.1 Followed steps to perform an indirect ELISA assay

Fig. 2.2 Example of a label-free sensor transducer. Antibodies (*red*) and antigen (*blue*) are immobilised on the surface of the sensor

labelling assays only provide endpoint read out, and do not allow continuous monitoring. It is thus not possible to provide information on the binding kinetics. Secondly, the multiple washing stages required between each step of the assay often complicates the required sample preparations, reducing the effective throughput and increasing the cost. And finally, the need to identify and produce two different antibodies that recognise different regions of the same target significantly increases the complexity of establishing a reliable assay.

Due to the above considerations, there has been a drive to develop label-free biosensors that reduce assay cost and complexity while providing quantitative information with high throughput. Label-free methods allow to continuously monitor the affinity reaction, providing highly quantitative measures of binding affinity and kinetics [4] and the variation in biomarker concentration over time. Label-free assays are typically surface-based, where the surface of a transducer is functionalised with a layer of receptor molecules. The assay development is also significantly simplified, particularly for highly multiplexed arrays, since only one recognition element is required for each analyte.

The sensor itself consists of a transducer, where the binding event causes a change in a physical property of the sensor which is subsequently measured (Fig. 2.2).

2.2 State-of-the-Art Label-Free Technologies

Here I provide a brief overview of some of the most prominent transduction technologies for label-free biosensing applications, while highly relevant publications are also deeply discussed in specific contexts within each chapter of this thesis.

Currently the most widespread and well-known example of label-free biosensor technology is surface plasmon resonance (SPR) [5], which directly measures the local refractive index change induced by biomolecular interactions at a gold surface using surface plasmon waves. Surface plasmons are charge density oscillations that can be excited optically. A number of commercial technologies based on this technique has already been developed by industrial companies such as Biacore (a division of GE

Healthcare), GWC Technologies, IBIS Technologies, Toyobo, GenOptics, SensiQ and Bio-rad. The best limits of detection (typically around 0.1 pg/mm^2 or ng/mL for protein interactions [5–8]) are obtained when interrogating the sensor with angular or wavelength spectroscopy, but then only up to ten measurements can be performed simultaneously [9], limiting the degree of multiplexing and throughput.

With the prospect of providing a robust and practical technology that can deliver limits of detection that rival that of SPR spectroscopy, while still allowing high degree of multiplexing with small fluidic sample consumption and high throughput, multiple technologies have emerged in the last decades.

For instance, sensors based on optical diffraction gratings can measure refractive index changes induced by molecular interactions through a shift of the diffraction wavelength in a fixed direction. They can be mass produced cost-effectively, and also commercial systems for high throughput cell-based assays and drug development are offered by SRU Biosystems and Corning. Nevertheless, simultaneous interrogation of a large number of wells of the microtiter plate in parallel requires optical imaging that only offers end-point read-out [10], instead of the real-time binding curves necessary for analysis of binding kinetics and affinity. Additionally, no highly multiplexed quantitative proteomics at clinically relevant limits of detection has been reported to date.

The potential advantages of integrated photonic sensors have also been exploided in the last decades. They can be easily miniaturised and they offer high potential for chip integration, while showing extremely high sensitivities (pM in a label-free scheme [7, 11, 12]). They also offer a high degree of flexibility in the materials and structures selection, while allowing the fabrication of arrays of sensors with the same characteristics within the same chip for multiplexing analysis. In the case of using silicon photonics technology, they provide additional advantages as are low power consumption and potential for mass production with subsequent reduction of production costs.

Electrochemical label-free techniques (Sect. 2.3.1), such as impedance biosensors and field-effect transistor sensors based on nanowires and carbon nanotubes, are well-suited for highly sensitive integrated systems (for instance, detection of fg/mL concentration values has been reported [13]). However, their performance often deteriorates at physiological ionic strengths (0.15 M), requiring desalting of the sample prior to the measurement [14].

2.3 Single-Domain Techniques

A number of strategies have been developed for transducing biomolecular binding between a surface immobilised probe molecule and a target biomarker in solution, including acoustic, electrical or optical sensing principles. Sensing systems which transduce the binding event via a single sensing mechanism can be classified as "single-domain" techniques. This section reviews the most established electrical and photonic waveguide-based single-domain sensing systems. In-depth reviews of these sensors can be found in [15, 16].

2.3.1 Electronic Biosensing

A large number of mechanisms can be exploited to transduce a biochemical signal to the electrical domain. Depending on how this transduction is done, a broad subdivision into six classes can be suggested: amperomic/voltammetric biosensors [17], potentiometric biosensors [18], conductometric biosensors [19], impedance biosensors [20] and field-effect transistor (FET) biosensors [21].

Amperomic/voltammetric biosensors measure the change in peak redox current (either oxidation or reduction) of a perturbed capture molecule immobilised on an electrode surface. Here, the capture molecule needs to be an enzyme for which the target is the corresponding enzyme substrate. A notable example is the electrochemical detection of glucose (diabetes biomaker) which employs the enzyme glucose oxidase. Since Clark proposed the first enzyme glucose biosensor in 1962 [22], the field received considerable attention which has lead to fast, sensitive (limits of detection down to 0.18 μM have been reported [23]) and reliable glucose biosensors [24].

Potentiometric biosensors exploit ion-selective electrodes whose potential responds selectively to the concentration of a given ion. Operated under conditions of negligible current flow, this electrode measures the accumulation of charge versus a reference electrode immersed in the analyte solution. In order to apply this principle to biosensing applications, for example the detection of proteins, one has to relate any change in the local concentration of ions at the electrode surface to the binding of the pointed molecules. This technique is capable of operation at extremely low detection limits (below ng/mL) as demonstrated in [25], although in this particular case the high sensitivity is achieved at the expense of throughput (10 h).

Conductance sensors employ a measurement of current flow (i.e. conductance) across a supporting solution that bridges two electrodes. To perform a test using this approach, one needs to link any change in the flow of charge through the substrate (e.g. a nanowire [26]) to the binding of a target to a capture molecule immobilised on the electrode surface. For example, a highly sensitive, conductometric label-free biosensor has been reported based on polyaniline nanowires between gold microelectrodes pairs on silicon [27]. The immobilisation of the immunoglobulin E aptamer onto the engineered nanowire enabled detection of immunoglobin at concentration as low as pg/mL [27].

The most widespread technique employed by the electrochemistry community to sensitively monitor variations on the resistivity/charging capacity of an electrochemical interface is the well-known Electrochemical Impedance Spectroscopy (EIS). This technique employs a small sinusoidal potential superimposed on a DC bias to an electrochemical cell whose impedance is measured as a function of the frequency. The complex impedance, which is defined as $Z(\omega) = X(\omega) + iY(\omega)$ where i equal to $\sqrt{-1}$, reveals information about the dielectric medium between the electrodes.

$Z(\omega)$ can be modelled as simple equivalent circuit where each component relates to different sections of the dielectric medium as depicted in Fig. 3.15.

When there is a redox related charge transfer across the electrode interface during measurement (Faradaic EIS), published assays employing immobilised surface antibodies are capable of sub ng/mL detection limits for a wide range of targets in aqueous solutions [28–31]. A simpler and potentially more suitable approach for practical applications is found with the absence of this redox probe (non-Faradaic EIS). Employing this technique, M. Dijksma and co-workers [13] were able to detect a biomarker for autoinflammatory and autoimmune disease with a sensitivity of 0.02 fg/mL.

Devices based on field-effect transistors (FETs) have also attracted great attention due to their ability to directly translate the interactions between targeted biological molecules and the FET surface into readable electrical signals [21]. In a standard FET, current flows along a semiconductor path (the channel) that is connected to two electrodes, (the source and the drain). The channel conductance between the source and the drain is switched on and off by a third (gate) electrode that is capacitively coupled through a thin dielectric layer. They can be sufficiently sensitive to achieve protein detection down to pg/mL or even lower levels [32].

Arrays of electrical biosensors have been demonstrated based on some of the detection strategies described previously. These arrays seek to find patient "fingerprints" through the quantification of several tens of makers [33]. For example, simultaneous and real-time detection of three cancer biomarkers has been demonstrated with the use of three antibody modified FET devices with concentrations down to pg/mL [14]. Table 2.1 provides a summary of the different levels of sensitivity obtained by the electrical biosensors discussed in this section.

In contrast to the electrical sensing platform, optical waveguide-based label-free biosensors allow for sophisticated and compact transducers due to the high confinement of light into the waveguide. These optical label-free biosensors evaluate changes in the propagation velocity of electromagnetic fields due to the presence of

Table 2.1 Summary of the different levels of sensitivity obtained by the electrical biosensors discussed in this section. PB stands for phosphate buffer solution, while PBS for phosphate buffer saline solution. All the sensitivity values reported here were reported without any amplification method

Sensor type	Analyte	Sensitivity	Analyte	Reference
Amperomic	Glucose	32.4 ng/mL	PB	[23, 24]
Potentiometric	Vascular endothelial growth factor	17.35 pg/mL	PB	[25]
Conductance	Immunoglobin	0.56 pg/mL	PBS	[27]
Non-faradaic EIS	Protein interferon gamma	0.02 fg/mL	PB	[13]
FET	Carcinoembryonic antigen (CEA)	0.5 ng/mL	PBS	[32]

biological particles. They translate changes in the propagation speed of light into a quantifiable signal proportional to the amount of biological material present on the sensor surface.

2.3.2 *Photonic Waveguide-Based Detection Strategies*

Sophisticated and compact label-free biosensors based on optical waveguides exploit the high confinement of light in a waveguide to transduce a biomolecular binding event. Such optical label-free biosensors evaluate changes in the propagation velocity of electromagnetic fields due to the presence of biological particles and translate changes in the propagation speed of light into a quantifiable signal proportional to the amount of biological material present on the sensor surface.

Recently, integrated photonic components based on silicon have become some of the most promising photonic integration platforms. This promise can be attributed to the combination of very high index contrast and the availability of CMOS fabrication technology, which enables the developments and facilities developed for the fabrication of microelectronics to be applied to Photonic Integrated Circuits (PICs) [34]. In this way, it is possible to design and optimise optical label-free biosensors exploiting the most recent progress in the PIC field.

2.3.2.1 Electromagnetic Waves

All the properties of electromagnetic waves, and their interaction with matter, are dictated by the four Maxwell equations [35]. In these equations, light is defined as an electromagnetic wave consisting of sinusoidally-varying electric (**E**) and magnetic (**B**) fields. Both fields have the same frequency (ν), are in phase and oscillate perpendicularly to each other. Light then propagates through space on a direction perpendicular to both **E** and **B** and is described by the orientation of the 'k-vector (**k**), such that **E**, **B** and **k** are all orthogonal for a plane wave. The 'wavenumber' ($|\mathbf{k}|$) can be understood as a spatial angular frequency: $k = \frac{2\pi}{\lambda}$, where λ is the wavelength, and the temporal angular frequency being $\omega = 2\pi\nu$. Therefore, the Maxwell equations describe the propagation of light in vacuum (where there are no sources or existing electrical charges or currents):

$$\nabla \cdot \mathbf{E} = 0 \quad (2.1)$$

$$\nabla \cdot \mathbf{B} = 0 \quad (2.2)$$

$$\nabla \times \mathbf{E} = -\frac{\partial \mathbf{B}}{\partial t} \quad (2.3)$$

$$\nabla \times \mathbf{B} = \mu_0 \epsilon_0 \frac{\partial \mathbf{E}}{\partial t} \tag{2.4}$$

where t, μ_0 and ϵ_0 represent the time and the vacuum permeability and permittivity, respectively. For the vacuum case, the two divergence Eqs. (2.1 and 2.2) state that there are no sources or sinks of **E** or **B** fields, while the two curl Eqs. (2.3 and 2.4) indicate that an oscillating magnetic field induces an oscillating electric field, while an oscillating electric field induces an oscillating magnetic field. Thus, due to these inductions between fields, electromagnetic waves do not require a medium in which to propagate. Maxwells equations are coupled first-order partial differential equations, but they can be decoupled to yield separate second-order equations for **E** and **B** [35]:

$$\nabla^2 \mathbf{E} = \mu_0 \epsilon_0 \frac{\partial^2 \mathbf{E}}{\partial t^2} \tag{2.5}$$

$$\nabla^2 \mathbf{B} = \mu_0 \epsilon_0 \frac{\partial^2 \mathbf{B}}{\partial t^2} \tag{2.6}$$

These take the form of equations describing a wave that has a propagation velocity of $\frac{1}{\sqrt{\mu_0 \epsilon_0}}$, which equals to speed of light in vacuum ($c \approx 3 \times 10^8 \mathrm{ms}^{-1}$). When an electromagnetic wave passes through a dielectric material, the speed at which light propagates is altered due to the change in the permeability and permittivity of the material (μ and ϵ, respectively), which is defined as:

$$\frac{1}{\sqrt{\mu \epsilon}} = \frac{c}{n}; n = \sqrt{\frac{\mu \epsilon}{\mu_0 \epsilon_0}} \tag{2.7}$$

being n the refractive index of the material, which represents how the propagation velocity is reduced compared to the vacuum velocity c. When the electromagnetic wave faces an interface with a material of different ϵ, the boundary conditions imposed by Maxwell's equations give rise to Snell's law of refraction and the Fresnel equations, which predict the reflection behaviour of light at a dielectric interface. Moreover, the time-space dependency in Maxwell's equations can be separated due to their linearity. Furthermore, by considering monochromatic harmonic waves (waves with a single frequency ω), they can be solved as:

$$\mathbf{E}(\mathbf{r}, t) = \mathbf{E}(\mathbf{r}) e^{-i\omega t} \tag{2.8}$$

$$\mathbf{B}(\mathbf{r}, t) = \mathbf{B}(\mathbf{r}) e^{-i\omega t} \tag{2.9}$$

The conformation of the spatial profile $\mathbf{E}(\mathbf{r})$ depends on the spatial distribution of ϵ, and the knowledge of this allows the spatial mode profiles (mode distribution) to be calculated by treating Maxwells equations as an eigenvalue problem. Following this approach, the resulting mode profiles are the eigenfunctions of the system, while the frequencies (ω) of the modes are the corresponding eigenvalues.

2.3.2.2 Silicon Waveguides

A waveguide is a structure that guides the propagation of waves, such as electromagnetic waves or sound waves. In this work, a waveguide guides light over the chip by total internal reflection (from Snell's law of refraction). It is based on a high refractive index core (n_2 in Fig. 2.3) surrounded by a low refractive index cladding (n_1 and n_3 in Fig. 2.3). Here, the high refractive index core is made of silicon while the cladding compromises both air and silicon dioxide.

The waveguide can guide multiple optical modes depending on whether the major electric field component is along the transverse direction (transverse electric modes, quasi-TE modes), or the major magnetic field component is along the transverse direction (transverse magnetic modes, quasi-TM modes) [36]. Each mode propagates across the waveguide with a phase velocity $\frac{c}{n_{eff}}$, where n_{eff} the effective refractive index perceived by the mode. As long as the effective refractive index is larger than the largest refractive index of the cladding (1.44 at 1.55 μm for silicon dioxide (SiO_2)), the mode is guided and confined in the waveguide. The higher the effective refractive index, the stronger the confinement.

Power can be transferred between multiple guided modes as light propagates through the waveguide, causing undesired interference effects that distort the transmission spectrum of optical components. To avoid such power transfer, waveguides need to be made sufficiently narrow to limit the number of modes [36]; typically, we use waveguides that only support a single mode for each polarisation. If the difference between the effective refractive index of the two remaining quasi-TE and quasi-TM modes is large, coupling between these modes will be limited because of the phase mismatch and very different mode profiles. For widths under 520 nm, the silicon photonic waveguide studied here will be single-mode for each polarisation [36].

The dimensions of these waveguides (typically around 500 × 220 nm) at the wavelength of 1.55 μm support the propagation of the quasi-TE mode (Fig. 2.4), which enables very compact and high-performance sensors as a result of this high confinement.

Fig. 2.3 Cross section profile of a waveguide on a substrate with air cladding. A high-refractive-index core (n_2) is surrounded by low-refractive-index media (n_1 and n_3) to confine the optical mode

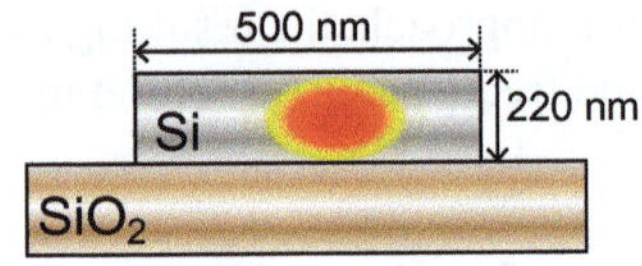

Fig. 2.4 Illustration of the high mode confinement in a standard waveguide for TE polarised light

Due to the fact that the propagation constant (n_{eff}) of an optical mode is obtained together with the mode profile as a solution of the Maxwells equations to the material system via an eigenvalue equation (Sect. 2.3.2.1), it is then conceived that this n_{eff} will change when the conditions at the waveguide are modified (as occurred when molecules bind to the surface of a waveguide). Perturbation theory states that a change in $\epsilon(\mathbf{r})$ leads to a change in the propagation constant relative to the size of the fields at the perturbation. Therefore, given that the fields at these perturbations exponentially decrease away from the core of the waveguide, these evanescent tails of the modes effectively sense the action close to the waveguide. This is, for most waveguide systems, a distance of the order $0.1 - 1\,\mu$m, and is known as the 'penetration depth'. This sensing phenomenon is called 'evanescence wave sensing', and can be evaluated by any of the waveguiding optical properties (polarisation, intensity, phase, resonance...). In Sect. 2.3.2.5, this sensing mechanism will be discussed in-depth for the sensors employed throughout this thesis.

Another important factor to consider is the waveguide dispersion. Dispersion means that the effective refractive index of a waveguide is strongly dependent on the wavelength. In ring resonator sensors, dispersion plays an important part as we will discuss. The group refractive index of an optical mode is a useful parameter as it takes first-order dispersion into account:

$$n_g = n_{eff} - \lambda \frac{dn_{eff}}{d\lambda} \tag{2.10}$$

with n_{eff} the effective refractive index and λ the wavelength in vacuum.

Here, all optical modes have a normal first-order dispersion, with a group refractive index that is larger than the effective refractive index. In particular, the group refractive index ($n_g = 4.35$) of the a quasi-TE mode waveguide, as used here, is nearly twice the effective refractive index ($n_{eff} = 2.33$), highlighting the importance of the dispersion.

2.3.2.3 Integrated Optical Label-Free Biosensors

Optical waveguides are able to transduce binding events (biochemical signals) into the optical regime; interferometric waveguide sensors (such as Mach-Zehnder interferometers [37] and bi-modal waveguide sensors [38]), photonic crystals (PhC) [39], slot-waveguides [40–42] and ring resonators [43] are examples of this capability. Initially, biosensing performed with integrated optical devices was mainly carried out on Silicon-On-Insulator (SOI) substrates. However, due to the strong water absorption at the operating wavelength of this silicon devices (which limits the performance of the sensing devices), integrated optical biosensing is nowadays primarily performed on silicon nitride based devices. Here I present a brief review of the biosensing performed with the most widespread and commonly employed devices, while I refer to [11, 12] for very comprehensive reviews of integrated optical label-free biosensors.

Mach-Zehnder interferometers (MZI) utilise the interference between two light-waves of equal wavelength where one of them has experienced a phase delay. Depending on the amount of phase delay (induced by the refractive index medium or optical path length), the optical waves will interfere constructively or destructively. Therefore, functionalising just one of these optical paths, binding events will change the refractive index locally and will induce a shift in the transmission spectrum. MZI have been shown to be able to detect refractive index changes of 10^{-7} RIU [37] and concentrations down to 1 fmol/μL using a silicon nitride substrate [44], while their use for the implementation of labonchip platforms has also been reported [45].

Contrary to MZI, bimodal waveguide interferometry sensors employ a single waveguide to perform the transduction of chemical events. In these sensors, two areas within the waveguide are employed: one operating in a single-mode regime, and a second one that supports two modes (zero- and first-order modes). As the evanescence field has a different distribution for each of the areas, they propagate at different velocities depending on the refractive index of the overlayer. In this way, the interference pattern at the exit of the waveguide changes if the refractive index varies, for instance, as a consequence of a biointeraction event. With this simple approach, based on a silicon nitride substrate, refractive index changes of 10^{-7} RIU [38] and concentrations down to 30 pg/mL of human growth hormone (HGH) [12] have been reported.

PhCs are basically structures with a periodic arrangement of material of different dielectric constant. The periodicity of the structure is very important as it defines the wavelengths that it can support. This periodicity can occur in 1D, 2D or 3D, yet we will discuss 2D structures which are the most widespread configurations. 2D structures have a varying dielectric function in two directions. Typically, triangular or square lattice of air holes are created in semiconductor materials such as silicon. When waves move through such periodic structures they scatter and interfere with themselves. When interference occurs, it results in the diffraction of the wave, forcing it to be sent out of the structure. When interference does not occur (waves that are physically supported by the periodic structure), waves continue to propagate through the structure. In such structures, light can be confined to high quality factor (Q-factor) cavities, directed, split and even slowed down. Their transmission spectrum contains a photonic bandgap in which almost all of the light is reflected for a wavelength range determined by the period of the lattice. Hence, they can be thought of as very efficient wavelength dependent mirrors, with frequencies within the bandgap being strongly reflected. Four detection strategies are used for the detection of biomarkers using PhCs: band-edge detection [46], defect based devices [47], resonant gratings [48] and angular spectrum sensing [49].

As an example of the high sensitivity detection which these devices can achieve, a defect-based PhC can detect 2.5 fg/mL of Bovine Serum Albumin (BSA) [50]. PhC can also be used for a wide range of interesting biosensing applications. For example, they can be placed on a fiber facet for in-situ sensing [51], used as a sensing device in an imaging method for detection of cancer cell cytotoxicity and proliferation [52] or for resonant surface imaging with high spatial resolution [53]. In the literature, we can find examples of optical biosensor arrays based on PhCs. A method to create large-scale chip-integrated photonic crystal biosensor arrays is demonstrated in [54]. In this example, four defect microcavities are able to sense binding events with 3.35 pg/mL concentrations.

Slot-waveguide sensors consist of two slabs of high refractive index substrates distanced by a nanometer-scale low refractive index slot region and surrounded by low refractive index cladding materials in such a way that light is strongly confined in the slot region. Due to this strong confinement, a stronger light-matter interaction can be achieved within this region as compared to conventional rib or planar waveguides, which results in an enhanced bulk sensitivity. Different slot-waveguide structures have been developed with different structures such as ring resonators [42], photonic crystals [41] and microdisk resonators [40]. However, despite the enhanced light-matter interaction, moderate sensitivity have been reported (changes of 10^{-6} RIU [41] and detection of 30 ng/mL [40]) due to the experimental noise of these systems which limits their sensitivity.

In summary, MZIs present very high sensitivity, however they require large footprints which is limiting for the fabrication of highly dense multiplexed arrays of sensors. PhCs also exhibit extremely high sensitivities due to the strong localisation of the light mode, nevertheless they are very sensitive to fabrication deviations. On the other hand, ring resonator based biosensors have also been extensively examined and also used in commercial products [55], which are more tolerant to fabrication deviations than MZIs and PhCs while still showing high quality resonant factors. This resonant structure, frequently used in integrated photonics, will support our work and is presented in the next section. Finally, a comparison between the different sensitivity levels obtained with the optical sensing techniques presented in this section is given in Table 2.2.

Table 2.2 Summary of the different levels of sensitivity obtained by the optical biosensors discussed in this section

Sensor type	Analyte	Limit of detection (RIU)	Sensitivity	Reference
MZI	DAPK gene	10^{-7}	≈1pg/mL	[44]
Bi-modal waveguide	HGH	10^{-7}	30 pg/mL	[38]
PhC	BSA	10^{-5}	12.5 fg/mL	[50]
PhC array	Avidin	10^{-4}	3.35 pg/mL	[54]
Slot-waveguide μdisk	Streptavidin	10^{-6}	30 ng/mL	[40]
Slot-waveguide PhC	Streptavidin	10^{-6}	15 ng/mL	[41]

2.3.2.4 SOI Ring Resonator Biosensors

Biosensors based on photonic ring resonators have been extensively examined and used in commercial products such as Genalyte [55]. Due to the fact that this resonant structure is robust to fabrication deviations and exhibit high quality resonant factors, it will be used extensively throughout this thesis.

In general, a ring resonator consists of a ring-waveguide and a coupling mechanism to access the loop. A mode propagating through the loop constructively interferes with itself when its wavelength fits an integer number of times in the loop. In other words, the cavity is in resonance when waves that travel in the loop have a round trip phase shift which equals an integer time 2π. Thus, the ring is in resonance at wavelengths λ_{res}:

$$\lambda_{res} = \frac{n_{eff} L}{m}, m \in \mathbb{N} \tag{2.11}$$

Here, n_{eff} is the effective refractive index of the resonant mode and L is the physical round trip length of the ring resonator. To provide access to its resonance modes, one or more access waveguide are placed next to the ring cavity. Depending on the number of these access waveguides, the whole configuration is called all-pass ring resonator (one access waveguide) or add-drop ring resonator (two access waveguides).

An all-pass ring resonator is a ring resonator in which a fraction k of the field that is propagating through its access waveguide is coupled into the cavity loop (Fig. 2.5). Without considering losses, the amplitude of this coupled wave is related to the coefficient $r = \sqrt{1 - k^2}$ [56].

The basic spectral properties of an all-pass ring resonator can be derived by assuming continuous wave regime and matching fields [56]. Assuming that reflections are negligible, L the physical round trip length of the ring cavity, a the single-pass amplitude transmission (including propagation losses in both the ring and in the couplers), it is possible to obtain the power transmission from squaring the ratio of the transmitted electrical field and the incident electrical field in the continuous wave regime:

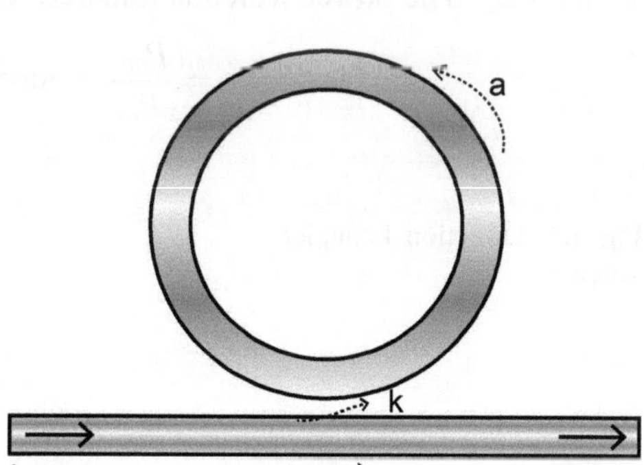

Fig. 2.5 All-pass ring configuration

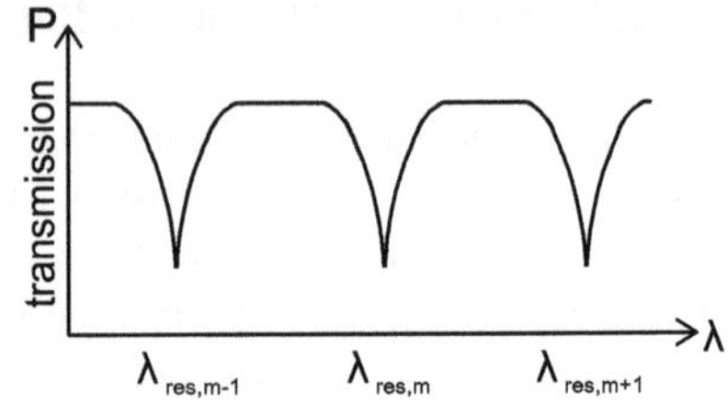

Fig. 2.6 Transmission spectrum of an all-pass ring resonator, showing dips at the resonance wavelengths

$$T_{allpass}(\lambda) = \frac{I_{pass}}{I_{input}} = \frac{a^2 - 2ar\cos(\frac{2\pi}{\lambda}n_{eff}(\lambda)L) + r^2}{1 - 2ar\cos(\frac{2\pi}{\lambda}n_{eff}(\lambda)L) + (ar)^2} \tag{2.12}$$

Under the assumption of a lossless ring ($a = 1$), the power transmission is unity for all wavelengths, hence the name *all-pass* ring resonator. However, in a practical case, where $a < 1$, thus the transmitted power presents minimum values at the resonance wavelengths (Fig. 2.6), due to destructive interference between the access wavelength mode and light coupled back into the waveguide from the ring.

Considering the relation between r and a, it can be seen that the ring can be overcoupled ($r < a$), undercoupled ($r > a$) or critically coupled ($r = a$). Critical coupling is an interesting regime for sensing as it allows accurate measurements of the resonance wavelength due to the sharpness that the resonance dip exhibits at this regime [56].

The coupling section, together with the access waveguide and the cavity make up the entire ring resonator. The access waveguide is brought close to the cavity so that the waveguide modes are coupled and power can be transferred from one to another [56]. This coupling is possible due to the portion of the electrical field (associated with the optical mode) that propagates outside the waveguide (commonly called evanescence field). When both waveguides are brought sufficiently close, the evanescence field can couple from the access waveguide into the cavity's waveguide (and vice-versa). Commonly, a directional coupler is employed to couple light from the ring's access waveguide to the ring's cavity and vice-versa.

As Fig. 2.7 shows, the directional coupler employed in the rings of our work consist of a section where two waveguides are close to each other with a gap G_c over a length L_c. The power which is transferred can be calculated as [57]:

$$k^2 = \frac{P_{out}}{P_{in}} = \sin^2(\kappa \cdot L_c + \kappa_0) \tag{2.13}$$

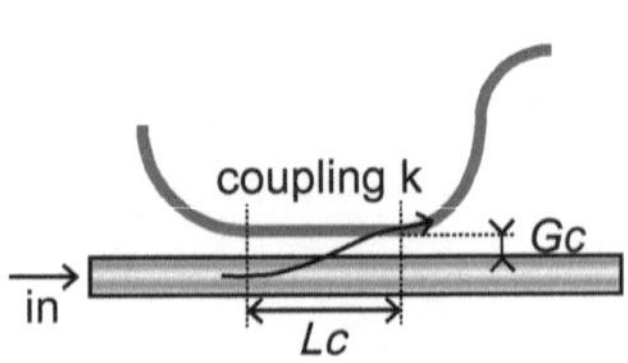

Fig. 2.7 Directional coupler scheme

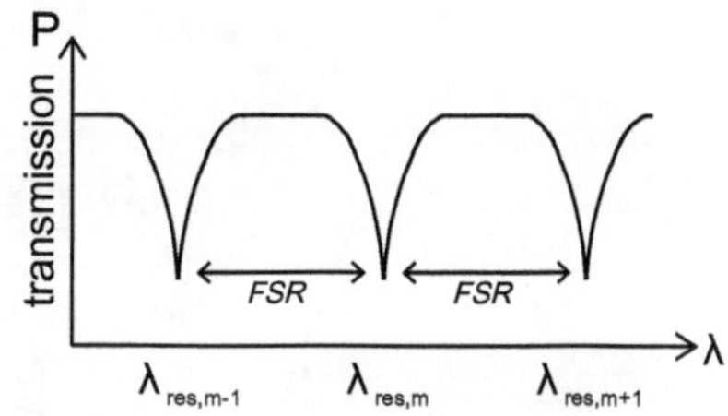

Fig. 2.8 Power transmission to the drop port of an all-pass ring resonator, where FSR indicates the spectral distance between subsequent resonances

where $\kappa[1/\mu m]$ is the coupling coefficient per unit distance in the closest section of the coupler and κ_0 is the offset coupling in this region. It is interesting to point out that the coupling strength (both κ and κ_0) increases with the wavelength, due to a decreasing mode confinement.

A key parameter for the all-pass configuration is the free spectral range (FSR). The FSR represents the spectral distance between subsequent resonances (Fig. 2.8), which as a function of wavelength equals:

$$FSR = \frac{\lambda^2}{n_g L} \tag{2.14}$$

where λ is the wavelength in vacuum, n_g is the group refractive index and L the physical round trip length of the ring's cavity. The transmission spectrum at the drop port of an all-pass ring resonator is depicted in Fig. 2.8, where the FSR is illustrated.

In photonic ring resonator biosensors, a larger FSR in comparison to the estimated $\Delta\lambda_{resonance}$ is desired in order to be able to accurately monitor the resonance wavelength along an experiment (i.e. FSR $>$ $\Delta\lambda_{resonance}$). In this way, interferences between adjacent resonances originated from large $\Delta\lambda_{resonance}$ are avoided when monitoring the resonance wavelength.

The performance of ring resonator sensors is mainly limited by losses [56]. Losses have a strong impact on the sharpness of the resonance leading to a reduced accuracy of the peak resonance position, due to the broadening of the resonance dip. Scattering due to sidewall roughness and absorption by the cladding layer are the major loss factors (≈3 dB/cm for air cladding and ≈4.7 dB/cm for water cladding [56]). Moreover, as has been mentioned, bend losses (≈0.04 dB/bend) also play an important role in ring resonator performance [56].

2.3.2.5 Evanescence Field Sensing with SOI Ring Resonators Biosensors

Biosensors based on ring resonators directly measure selective affinity interactions between analyte molecules in solution and receptor molecules immobilised on the ring waveguide surface (Fig. 2.9).

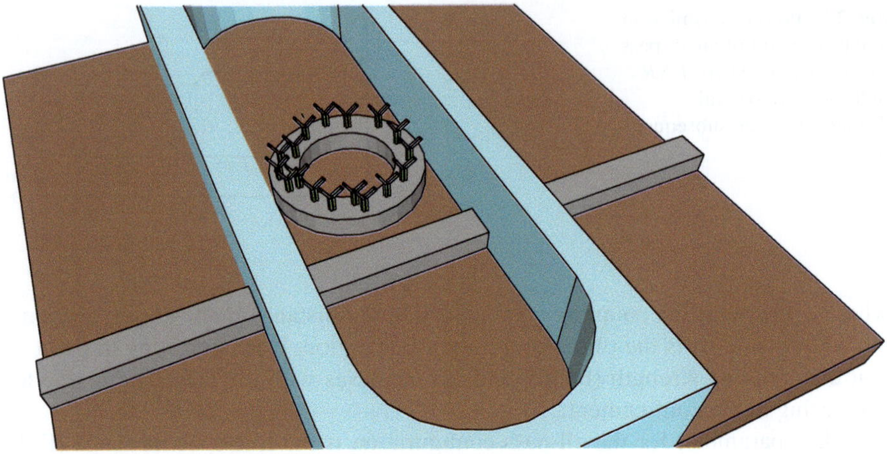

Fig. 2.9 Ring resonator biosensor coated with immobilised receptor molecules

Fig. 2.10 Spectral shift of a ring resonator biosensor due to a change in the effective refractive index

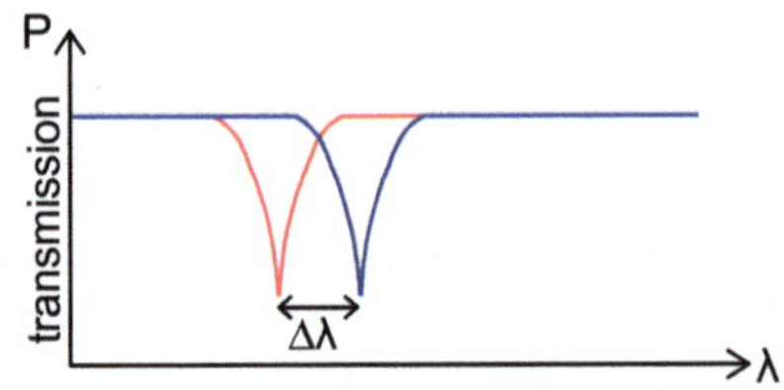

Since nearly all biological molecules have a larger refractive index than the surrounding aqueous solvent(for instance, $n_{antibodies} \approx 1.45$ Refractive Index Unit (RIU) and $n_{water} \approx 1.31$ RIU), this molecular binding increases the local refractive index in the area where the tail of the evanescent field of the waveguide mode is present. The consequent phase change of the waveguide mode produces a measurable shift of the resonance wavelength ($\Delta\lambda$) of the ring (Fig. 2.10). This shift can be monitored to give detailed information about the analyte concentration, affinity between the molecules and the kinetics of the biochemical reaction [57].

The evanescent field of a guided mode is the fraction of the mode that extends in the cladding (out of the waveguide). It decays exponentially with the distance from the core cladding interface [58]:

$$\left|\vec{E}(d)\right| = \left|\vec{E}(0)\right| \cdot \exp(-\gamma d) \tag{2.15}$$

where $\left|\vec{E}(d)\right|$ is the electric field in the cladding as a function of the distance d from the interface. γ is the decay constant that is formulated as [59]:

$$\gamma = \frac{2\pi}{\lambda}\sqrt{n_{eff}^2 - n_w^2} \tag{2.16}$$

Here, n_{eff} is the effective refractive index of the waveguide mode and n_w the refractive index of the cladding. Thus, a large refractive index contrast results in a short decay length. This is advantageous when sensing the presence of a thin layer of biomolecules as the mode only will interact with changes in the cladding that happen a few nanometers from the waveguide's surface in the region where the biomolecules are immobilised. In this way, the sensor is very sensitive towards refractive index changes caused by molecular binding near that surface. The greater the overlap between the immobilised molecular layer and the evanescent tail, the more sensitive the sensor. Furthermore, as the sensing zone remains very close to the waveguide's surface, the measurement will be less disturbed by the solution away the surface (improving signal-to-noise ratio). The length of the evanescent tail in solution can be tuned by modifying the geometry of the sensor. This will be exploited in Chap. 6 to quantify conformational changes of immobilised molecules on the surface.

If the refractive index of the cladding changes, the effective refractive index of the waveguide mode will also change (in the range of the waveguide mode evanescent field). Then, the resonance wavelength of the ring resonator mode will shift following the expression [43]:

$$\Delta\lambda_{resonance} = \frac{\lambda_{resonance} \cdot \Delta n_{eff}}{n_g} \tag{2.17}$$

where n_g is the group index that takes into account the large first order dispersion of the waveguide (Eq. 2.10).

Therefore, the sensitivity for ring resonators is defined as the shift of the resonance wavelength for a certain excitation. For sensing the refractive index of the fluid flowing over the sensor's surface, bulk sensitivity is a more useful metric. Its units are [nm (resonance shift)/RIU]. Another useful metric for evanescence field sensing is the surface sensitivity, which corresponds to the sensitivity of the device upon surface-related molecular interactions. This metric, usually given in [nm (resonance shift)/nm (layer thickness)], provides a quantification of the molecular mass bound to the sensor surface and can help to compare non-optical different technologies with optical ones.

Another important concept related to the sensor's sensitivity is the smallest wavelength shift that can be measured, $\Delta\lambda_{min}$, which is associated with the precision of the measurement equipment and the sharpness of the resonance wavelength relative to its central frequency, the quality factor (Q-factor).

Combining these two concepts, it is possible to define the limit of detection (LOD) of a sensor. The LOD refers to the minimum detectable concentration of an analyte in a test sample, and follows:

$$LOD = \frac{\Delta\lambda_{min}}{sensitivity} \tag{2.18}$$

The high Q-factor associated with ring resonators, can result in biosensors with sensitivities comparable to SPR [60]. Detection limits on surfaces of 1.5 pg/mm^2 (corresponding to 7.6×10^{-7} RIU) have been reported with a multiplexed sensing scheme [61], while cortisol-bovine serum albumin binding has been detected at protein concentrations in solution down to 1 ng/mL [62]. In a more sophisticated example, C. F. Carlborg et al. presented a fully packaged optical device based on slot-waveguide ring resonators sensors [63]. This example employed a variety of ring resonators where the optical signal is confined within two Si waveguides, hence sensing the analyte with the bulk optical signal and not with the evanescence tail. Here, a detection limit of 0.9 pg/mm^2 was demonstrated.

2.4 Bi-Domain Techniques

While single-domain techniques have been demonstrated to provide accurate, fast and reliable analysis of chemical events at the surface of the sensor, techniques that monitor these chemical events in two measurement domains can provide further insight into the undergoing molecular and (bio)molecular processes. For this reason, they are readily becoming more popular within the scientific community. For example, dual-mode (or bi-domain) techniques can simultaneously probe both the electrochemical and optical properties of thin films on the surface at the nanometer scale [64], or the viscoelastic properties and conformational state of deposited macromolecules [65]. The number of solutions developed to date for dual-mode (or multi-mode) sensing is significantly more limited than that available for single-domain techniques. Here we highlight the major dual-mode sensing strategies and some of their most relevant applications, while a comparison between them and the system proposed in this thesis is given in Sect. 4.8.

2.4.1 Electrochemical Surface Plasmon Resonance

SPR has significantly evolved since it was first applied for sensing three decades ago. A range of sub-techniques have also emerged from this sensing technique, including Localised Surface Plasmon Resonance (LSPR) [66], Long-Range Surface Plasmon Resonance (LR-SPR) [67] or Metal-Clad WaveGuide (MCWG) [68], which have been proven to be suitable for performing sensing and imaging. Of particular importance here is the simultaneous characterisation of the optical and electrochemical properties of immobilised molecules on the surface achieved with electrochemical-SPR (EC-SPR). For example, using EC-SPR, characterization of electrochemical DNA sensors has been reported [69], capable of detection of concentrations down to 20 nM. As discussed in Sect. 2.1, SPR sensors directly measure the local refractive index change induced by biomolecular interactions on a gold surface using surface plasmon waves. The transducer translates chemical changes into changes in refractive index, which may be determined by optically interrogating the SPR. The sensor sensitivity, stability, and resolution depend upon properties of both the optical system and the transducing medium. SPR allows the incorporation of a secondary sensing

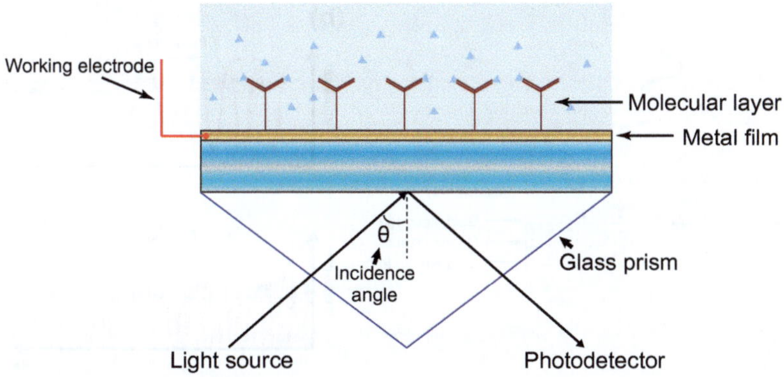

Fig. 2.11 Electrochemical-SPR schematic diagram with a three electrode cell set-up. The gold substrate that carries the optical surface mode is simultaneously used as the working electrode, allowing simultaneous optical and electrochemical interrogation

domain, here the electrochemical, where the gold substrate that carries the optical surface mode is simultaneously used as the working electrode in a standard three-electrode electrochemical system (Fig. 2.11).

This combination of electrochemical and optical sensing modes can provide deeper insight into redox active system, for example enzymatic reactions [64]. Nevertheless, the EC-SPR technique presents some disadvantages. Firstly, SPR requires a stable optical setup requiring bulky equipment to perform accurate and repeatable measurements. As a consequence, this technique is not a compact solution and thus limits application in point-of-care devices which can be used at the bedside. Secondly, high density arrays of SPR sensors are particularly difficult to engineer.

2.4.2 Electrochemical Quartz Crystal Microbalance with Dissipation Monitoring

A rapidly emerging label-free biosensor that enables picomolar level detection alongside simultaneous measurements of conformational changes in real-time is Quartz Crystal Microbalance with Dissipation monitoring (QCM-D).

QCM-D relies on a voltage being applied to a piezo electric quartz crystal causing it to oscillate at a specific resonant frequency (Fig. 2.12a). Changes in mass on the quartz surface, for example due to protein binding, result in a change of the resonant frequency of the oscillating crystal. The Sauerbrey relationship (method for correlating changes in the oscillation frequency of a piezoelectric crystal with the mass deposited on it) defines the relationship between mass and resonant frequency for rigid, homogeneous and thin adsorbed layers; the model is not valid for soft or viscoelastic films. In addition to measuring changes in resonant frequency, QCM-D

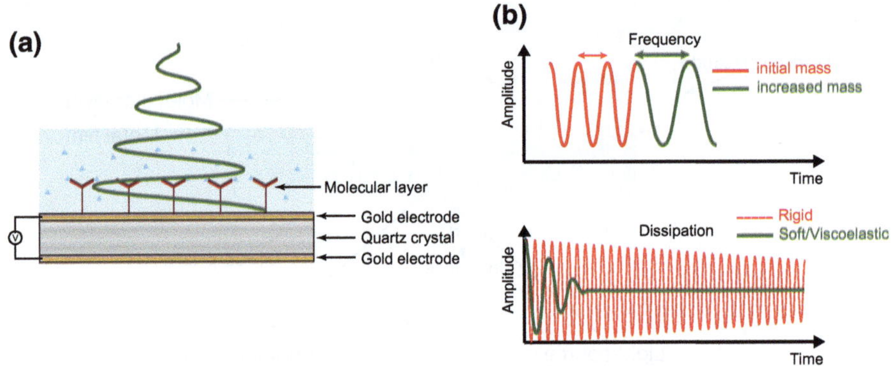

Fig. 2.12 Electrochemical Quartz crystal Microbalance with Dissipation monitoring (E-QCM-D). **a** The sensor crystal can be coated with a conductive metal layer, which is simultaneously used as the working electrode, allowing simultaneous mechanical and electrochemical interrogation. **b** Both the resonant frequency and the energy dissipation are measured simultaneously for a non-driven sensor crystal

also measures the energy dissipated by the acoustic wave [65]. This is achieved by disconnecting the drive voltage from the crystal while recording the time taken for the oscillation to decay due to damping (Fig. 2.12b). In this system, electrochemical control over the surface of the sensor can also be achieved by coating the quartz sensor with a thin metal layer which acts as the working electrode in a standard three-electrode electrochemical system.

Changes in adsorbed mass of, for example, a rigid protein provide a change in frequency, however for viscoelastic masses such as biomacromolecules, there is also a corresponding change in dissipation, leading to a rapid oscillatory decay of the stored energy in the crystal (Fig. 2.12b). This change in dissipation provides information regarding the viscoelastic properties of the immobilised layer and can thus be used to probe molecular conformation.

Additionally, QCMD allows simultaneous measurements of frequency shift and energy dissipation of multiple different overtones of the fundamental resonance. As higher overtones are more confined into the quartz crystal, while lower overtones are less confined and penetrate further into the supporting analyte, it is possible to also probe the distribution of mass and viscoelasticity as a function of distance into the immobilised layer.

E-QCM-D is an ideal technique for detection of mass on surfaces, as any change is mass on the sensor surface is directly translated onto the resonant frequency. In contrast with optical techniques, E-QCM-D is sensitive to water associated with adsorbed proteins, allowing the quantification of hydrodynamically coupled water (proteins and/or water trapped in cavities in the film).

This electro-mechanical system is also well suited for the detection of enzymatic activities. For instance, a 21 pM threshold of enzymatic activity has been reported while being able to distinguish different hydrolysis mechanisms [70].

2.4.3 *Electrochemical Optical Waveguide Lightmode Spectroscopy*

Another (however less developed) optical dual-mode experimental technique has been developed called Electrochemical Optical Waveguide Lightmode Spectroscopy (EC-OCWLS). EC-OWLS combines evanescent-field optical sensing with electrochemical control of surface adsorption processes.

This technique, firstly presented in 2012, is based on grating-assisted coupling of light into an optical waveguide layer [71]. The angle of the incident light beam is then varied, exciting both guided transverse electric (TE) and transverse magnetic (TM) modes. As the optical mode penetrates into the supporting solution for a distance of about 200 nm beyond the surface, the in-coupling angle of the incident light is sensitive to the refractive index and the thickness of the adsorbed film on the surface. A layer of indium tin oxide (ITO) is deposited on top of the waveguide and serves as both a high-refractive-index waveguide and conductive electrode, being completely compatible with the constraints of optical sensing (Fig. 2.13).

EC-OCWLS can further be exploited to calculate the refractive index of the layer deposited on the ITO surface. This calculation is performed by measuring both TE and TM modes, as detailed in [72]. Alternatively, EC-OWLS has been reported to be able to measure adsorbed mass down to 0.5% of an average protein layer, showing its capabilities for studying protein adsorption kinetics [73].

2.4.4 *Electro-Photonic Silicon Biosensing*

It has become clear that with the introduction of the personalised medicine (PM) [74], portable, compact and reliable sensing devices will be required in the near future. For example, particular diseases, such as cancer, will require detection of biomarker profiles to personalise diagnosis and treatment. The SOI platform is, therefore, conveniently positioned to address this challenge presented by PM, as its CMOS com-

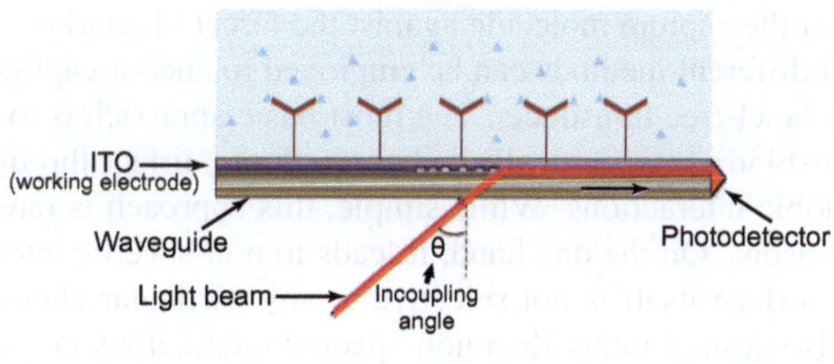

Fig. 2.13 Electrochemical Optical Waveguide Lightmode Spectroscopy (EC-OCWLS). Light coupled into the waveguide through a grating coupler changes its in-coupling angle when the refractive index on top of the sensor surface is modified

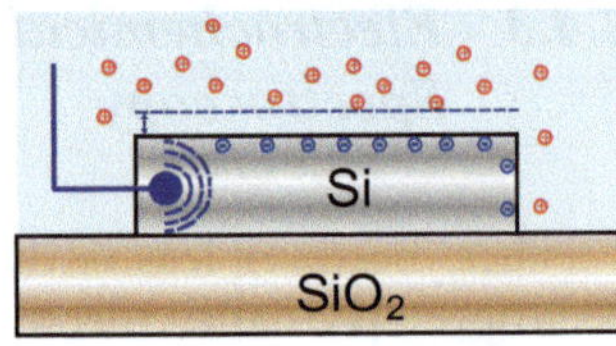

Fig. 2.14 Electro-photonic silicon biosensing. Electrochemical control to the surface of a SOI sensor is provided, allowing optical evanescence field sensing

patibility allows the development of such cheap and precise point-of-care devices which will be able to monitor multiple biomarkers in parallel.

Nevertheless, this platform can also be complemented by performing parallel sensing in a complementary domain, here the electrochemical (Fig. 2.14). In this way, information about the electrochemical activity of immobilised molecules, and deeper insight into molecular and (bio)molecular processes, can be provided in addition to that supplied by the optical domain. This novel sensing system, namely electrochemical-photonic silicon biosensing, is the object of study of this thesis.

To successfully combine both the electrochemical and optical domains, an electrochemically compatible layer has to be integrated within the SOI platform. For instance, an electrochemically and optically compatible layer could be deposited on the silicon layer of the SOI substrate or the electrical properties of the SOI substrate itself can be modified. The viability of both approaches is studied in Sect. 4.1.

This bi-domain system would take advantage of the SOI sensor platform (i.e. freedom to choose the optical structure for sensing, tailoring of light-matter interaction and read-out integration) while simultaneously conducting (and monitoring) electrochemical reactions occurring on the silicon surface. This not only provides insight into electrochemical processes but can also be exploited for chemical modification of the photonic sensors.

2.5 Self-assembled Monolayers

A critical component of all label-free technologies is the need to immobilise capture molecules onto the sensor surface using a process that must preserve the high affinity and selectivity of the capture molecule against the target biomarker.

A number of different methods can be employed to anchor capture molecules to the surface of a label-free transducer. The most basic approach is to allow the capture molecules to bind non-specifically to the transducer surface through electrostatic and/or hydrophobic interactions. While simple, this approach is rarely used due to the following reasons. On the one hand, it leads to non-specific interactions on the surface, as the surface itself is not selective to any particular chemistry. Covalent attachment is also desired rather than non-specific interactions in order immobilise the molecules on the surface. Furthermore, non-specific interactions do not allow to control the surface density or, when possible, the orientation of immobilised molecules. This is very important in biosensing as the need to optimise the bioassay conditions for the detection of molecules. Controlling the surface density allows to prevent steric hindrance, effect which can limit the performance of biosensors [75]. Finally, when proteins are used as capture molecules, their stability is affected when their are brought into contact with inorganic material, as they undergo a change in their structure (generally unfolding) with the loss of activity [76].

A more controlled approach to immobilisation is to pre-functionalise the surface in order to control the surface chemistry. This functionalisation, which can be patterned in different ways [77], introduces chemical functional groups to the surface of the transducer which can further link to other molecules through covalent chemical bonding.

The most widely utilised method of surface functionalisation is based on Self-Assembled Monolayers (SAMs). SAMs have received significant interest due to their ability to modify and control the chemistry and properties of planar surfaces, including semiconductors and metals, and curved surfaces, such as nanoparticles, as they form a well-ordered and stable thin film. SAMs are molecular assemblies spontaneously formed on surfaces by adsorption of a surfactant on a solid surface. Note that adsorption refers to the adhesion of atoms, ions, or molecules from a gas, liquid, or dissolved solid to a surface, while absorption relates a physical or a process in which atoms, molecules or ions enter some bulk phase. In contrast with ordinary surfactant monolayers, the molecules which form the SAM typically possess a chemical moiety (the head group as depicted in Fig. 2.15), that has a strong affinity for the substrate and thus tethers the molecule stably to the surface to expose the tail group (which acts as a physical barrier and provides a well-defined thickness). These functional tail groups are typically chemically functional, for example thiols (-SH), hydroxyls (-OH) and amines (-NH_2), allowing the covalent coupling of receptor molecules, as required for the design of label-free biosensor. In addition to chemical functionality for molecular immobilisation, SAMs can also provide further control of surface properties including chemical resistance, biocompatibility, wetting and adhesion [78]. The most widely investigated SAMs are those based on monolayers of alkanethiols on gold substrates [79].

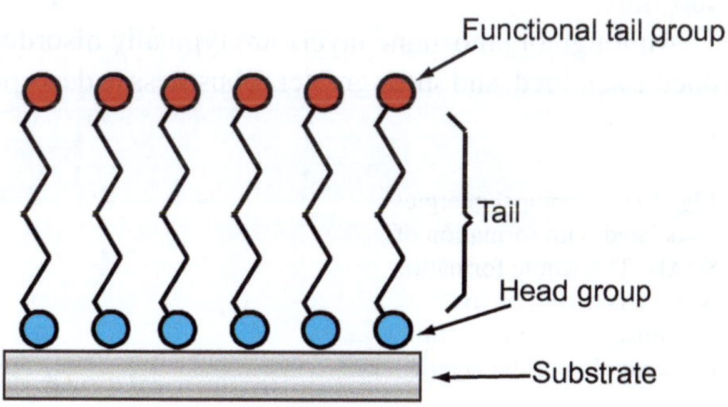

Fig. 2.15 Schematic of a SAM on a substrate. The "head group" is chemically absorbed onto the substrate, followed by a slow rearrangement of "tail groups" driven by intermolecular interactions between the organic tail region

2.5.1 Formation of Self-assembled Monolayers

SAMs are created by the chemisorption of head groups onto the substrate from either liquid of vapour phase, followed by a slow rearrangement of “tail groups”. The formation process is driven by the minimisation of energy in the formation of the layer, as adsorption lowers the surface free-energy of the system [78]. The main energies involved in the assembly of the monolayer are four: the energy of adsorption of the headgroup to the surface ($\triangle E_{ads}$), the energy related to the corrugation of the surface ($\triangle E_c$), the Van de Waals energies between the backbones ($\triangle E_v$) and the energy associated with the conformational isomers of adjacent groups along the backbone (gauche conformation in the case of alkane chains) ($\triangle E_g$). These energies are depicted in Fig. 2.16.

A key aspect for the quality of the SAM is the surface topography of the substrate. Surface roughness on the scale of the molecule will have a direct impact on the SAM quality, as the integrity of the SAM is inversely proportional to the roughness dimensions. Indeed, the quality of the SAM is more associated to the amount of grains in the substrate than to the amount of adsorbed molecules which construct the SAM [78]. However, if optimised, it has been reported that the roughness of the SAM can be employed to reduce non-specific adsorption in microfluidic biosensors [80].

2.5.2 Assembly of Silane SAMs on Silicon

Alkoxysilanes are some of the most common chemistries employed for the assembly of SAMs on both silicon and glass substrates through a process known as silanisation. Silanisation requires substrates to contain hydroxyl groups on the surface in order to form a covalent Si-O-Si bond between the surface and the alkoxysilane [78]. An example for the formation of a 3-Mercaptopropyltrimethoxysilane (MPTS) layer on silicon is shown in Fig. 2.17, which provides a thiol modification to the silicon substrate.

Although organosilane layers are typically disordered, they exhibit high stability once assembled, and show greater robustness to desorption than alkanethiol monolay-

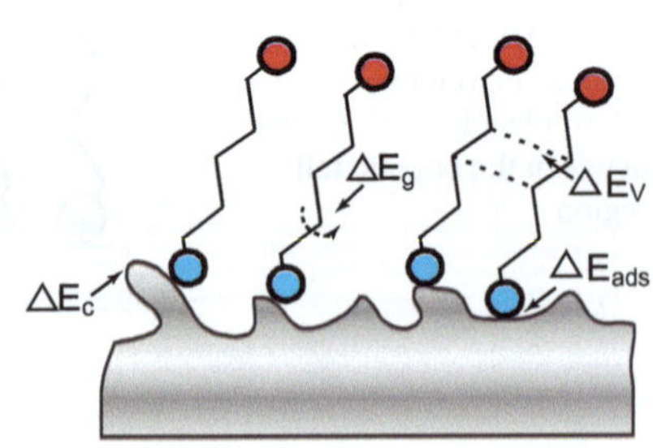

Fig. 2.16 Principal energies associated with formation of SAMs. The whole formation process is driven by the minimisation of energy in the formation of the layer

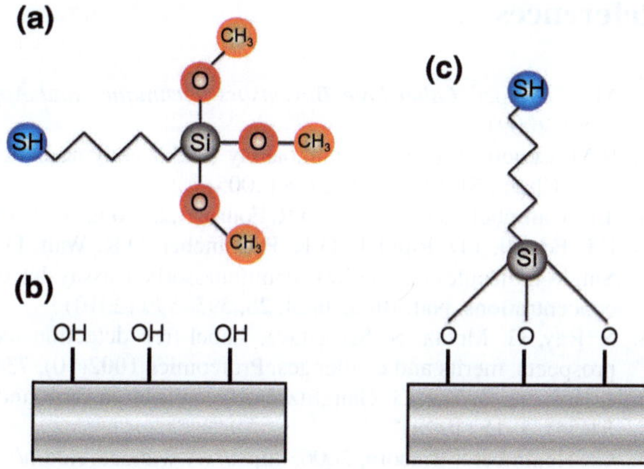

Fig. 2.17 Formation of a 3-Mercaptopropyltrimethoxysilane (MPTS) layer on silicon. **a** 3-Mercaptopropyltrimethoxysilane molecule requires hydroxil groups on the surface **b** to form a covalent Si-O-Si bond to the surface **c**

ers formed on gold [81]. Improvements to the order and reproducibility of organosilane layers requires minimisation of water content (both in the solution and on the surface). Excess presence of water molecules in the solution promotes the formation of polysiloxanes in the solution-phase and the assembly of multilayers on the surface [82], which decreases the quality of the layer.

2.6 Summary

This chapter has introduced the main motivation for the development of label-free biosensors, as the need of biofunctionalisationing such label-free transducers for the detection of biomolecules in their native conformation. The most widespread label-free detection strategies in both electrical and photonic domains have been highlighted, providing a summary of their sensing principles and main applications. More specifically, photonic waveguide-based detection strategies, and in particular ring resonator biosensors, have been the focus. The evanescence field sensing principle of such resonators has been extensively presented, as the devices here developed base their optical sensing mechanism in this effect. Subsequently, the chapter has followed with an introduction to bi-domain sensing techniques. It has been shown how the added value of these label-free dual-mode systems will help to understand complex molecular and biomolecular processes. The chapter follows introducing the highly novel dual-mode electrochemical-photonic sensing technique developed in this thesis, which is capable of, for instance, exploiting chemical modification of the photonic sensors, monitoring in situ electrochemical reactions occurring on the silicon surface or enzyme activity in parallel with substrate binding. Finally, a summary of the most common functionalisation techniques for silicon surfaces based on SAMs is given.

References

1. M.A. Cooper, *Label-Free Biosensors Techniques and Applications* (Cambridge University Press, 2009)
2. R.M. Lequin, Enzyme immunoassay (EIA)/ enzyme-linked immunosorbent assay (ELISA). Clin. Chem. **51**(12), 2415–2418 (2005)
3. T.G. Campbell, S.C. Howes, D.R. Fournier, L. Song, T. Piech, P.P. Patel, L. Chang, A.J. Rivnak, E.P. Ferrell, J.D. Randall, G.K. Provuncher, D.R. Walt, D.M. Rissin, C.W. Kan, D.C. Duffy, Single-molecule enzyme-linked immunosorbent assay detects serum proteins at subfemtomolar concentrations. Nat. Biotechnol. **28**, 595–599 (2010)
4. S. Ray, G. Mehta, S. Srivastava, Label-free detection techniques for protein microarrays: prospects, merits and challenges. Proteomics **1002**(10), 731–748 (2009)
5. J. Homola, S. Yee, G. Gauglitz, Surface plasmon resonance sensors: review. Sens. Actuators **54**(54), 3–15 (1999)
6. GE Healhtcare. Biacore, 2006, https://www.biacore.com/lifesciences/index.html
7. M. Alvarez, M. Carmen, Estevez, L.M. Lechuga, Integrated optical devices for lab-on-a-chip biosensing applications. Laser Photon. Rev. **6**(4), 463487 (2012)
8. J.M. Rodrguez-Frade, M. Mellado, J. Trevio, A. Calle, and L.M. Lechuga, Determination of human growth hormone in human serum samples by surface plasmon resonance immunoassay. Talanta **78**(3), 1011–1016 (2009)
9. J. Homola, Surface plasmon resonance sensors for detection of chemical and biological species. Chem. Rev. **108**(2), 462–493 (2008)
10. B.T. Cunningham, L. Laing, Microplate-based, label-free detection of biomolecular interactions: applications in proteomics. Expert Rev. Proteomics **3**(3), 271–281 (2006)
11. R. Puchades, M.-J. Bauls, N. Maquieira, Chemical surface modifications for the development of silicon-based label-free integrated optical (io) biosensors: a review. Analytica Chimica Acta **777**, 1–16 (2013)
12. J.C. Ramirez A. Fernndez Gavela, D. Grajales Garca and L. M. Lechuga. Last advances in silicon-based optical biosensors. Sensors **16**(3):285–300, 2016
13. J.C. Hoogvliet, M. Dijksma, B. Kamp, W.P. van Bennekom. Development of an electrochemical immunosensor for direct detection of interferon-gamma at the attomolar level. Anal Chem. **5**(73), 901–907 (2001)
14. Y. Cui, W.U. Wang, G.F. Zheng, F. Patolsky, C.M. Lieber, Multiplexed electrical detection of cancer markers with nanowire sensor arrays. Nat. Biotechnol. **23**(10), 1294–1301 (2005)
15. Florian Reuter Martin Nirschl, Janos Vrs, Review of transducer principles for label-free biomolecular interaction analysis. Biosensors **1**, 70–92 (2011)
16. A.P.F. Turner, Biosensors: sense and sensibility. Chem. Soc. Rev. **42**, 3184–3196 (2013)
17. X. Luo, J. Davis, Electrical biosensors and the label free detection of protein disease biomarkers. Chem. Soc. Rev., 5944–5962 (2013)
18. S. Liu, X. Guo, Carbon nanomaterials field-effect-transistor-based biosensors. Npg Asia Mater., 121–131 (2012)
19. J.H. Niazi Y. Gurbuz B.S. Youn, J.W. Park, S.S. Kallempudi, M.B. Gu, Rapid and sensitive detection of nampt (pbef/visfatin) in human serum using an ssdna aptamer-based capacitive biosensor. Biosens. Bioelectron., 233–238 (2012)
20. W. Lorenz, K.D. Schulze, Zur anwendung der transformationsimpedanzspektrometrie. J. Electroanal. Chem. Interfacial Electrochem. (1975)
21. P. Bergvled, The development and application of fet-based biosensors. Biosensors, 15–33 (1986)
22. L.C. Clark, C. Lyons, Electrode systems for continuous monitoring in cardiovascular surgery. Ann. New York Acad. Sci. **102**(1), 29–45 (1962)
23. J. Wu, L. Yin, Platinum nanoparticle modified polyaniline-functionalized boron nitride nanotubes for amperometric glucose enzyme biosensor. ACS Appl. Mater. Interfaces **3**(11), 4354–4362 (2011)

24. Q. Xie, D. Yang, H. Xiao, Y. Fu, Y. Tan, C. Chen, S. Yao, Recent advances in electrochemical glucose biosensors: a review. RSC Adv. **3**, 4473–4491 (2013)
25. L. G. Bachas K. M. L. May, A. Vogt, K. W. Anderson. Vascular endothelial growth factor as a biomarker for the early detection of cancer using a whole cell-based biosensor. Anal. Bioanal. Chem., 1010–1116 (2005)
26. B. Li, K.I. Chena, Y. Chena, The development and application of fet-based silicon nanowire field-effect transistor-based biosensors for biomedical diagnosis and cellular recording investigation. Nano Today, 131–154 (2011)
27. J.Y. Huang, M.H. Yun, X.L. Luo, I. Lee, X.Y.T. Cui, Ultrasensitive protein detection using an aptamer-functionalized single polyaniline nanowire. Chem. Commun. **47**(12), 6368–6370 (2011)
28. X.L. Luo, M.Y. Xu, J.J. Davis, The label free picomolar detection of insulin in blood serum. Biosens. Bioelectron. **39**(1), 21–25 (2013)
29. H. Rahmani, B. Rezaei, N. Majidi, T. Khayamian, Electrochemical impedimetric immunosensor for insulin like growth factor-1 using specific monoclonal antibody-nanogold modified electrode. Biosens. Bioelectron. **26**(5), 2130–2134 (2011)
30. A. Abulrob, A.C. Tavares, R. Elshafey, C. Tlili, M. Zourob, Label-free impedimetric immunosensor for ultrasensitive detection of cancer marker murine double minute 2 in brain tissue. Biosens. Bioelectron. **39**(1), 220–225 (2013)
31. N. Zine, G. Gabriel, A. Guimera, F. J. del Campo, R. Villa, A. H. Eisenberg, M. Mrksich, A. Errachid, J. Aguilo, E. Prats-Alfonso, X. Sisquella, F. Albericio, Cancer prognostics by direct detection of p53-antibodies on gold surfaces by impedance measurements. Small **8**(1), 2106–2115 (2012)
32. B.R. Li, K.I. Chen, Y.T. Chen, Silicon nanowire field-effect transistor-based biosensors for biomedical diagnosis and cellular recording investigation. Nano today **6**(2), 131–154 (2011)
33. A.O. Aluoch, O.A. Sadik, A.L. Zhou, Status of biomolecular recognition using electrochemical techniques. Biosens. Bioelectron. **24**(9), 2749–2765 (2009)
34. W. Bogaerts, R. Baets, P. Dumon, V. Wiaux, S. Beckx, D. Taillaert, B. Luyssaert, J. Van Campenhout, P. Bienstman, D. Van Thourhout, Nanophotonic waveguides in silicon-on-insulator fabricated with cmos technology. J. Lightwave Technol. **23**(1), 401–412 (2005)
35. D.J. Griffiths, *Introduction to Electrodynamics*, 3rd edn. (Prentice-Hall International Inc., NJ 07458, 1999)
36. K. Okamoto. *Fundamentals of Optical Waveguides*(Elsevier, 2006)
37. R.G. Heideman, P.V. Lambeck, Remote opto-chemical sensing with extreme sensitivity: design, fabrication and performance of a pigtailed integrated optical phase-modulated machzehnder interferometer system. Sens. Actuators B: Chem. **61**(13), 100 – 127 (1999), http://www.sciencedirect.com/science/article/pii/S092540059900283X
38. C. Dominguez, K.E. Zinoviev, A.B. Gonzalez-Guerrero, L.M. Lechuga, Integrated bimodal waveguide interferometric biosensor for label-free analysis. J. Lightwave Technol. **13**(29) (2011)
39. Y.A. Vlasov, S. Assefa, C. Kang, C.T. Phare, S.M. Weiss, Photonic crystal slab sensor with enhanced surface area. Opt. Express **18**(26), 27930–27937 (2010a)
40. S. Lee, S. Chan, Eom, J. Soo Chang, C. Huh, G. Yong Sung, J. H. Shin, Label-free optical biosensing using a horizontal air-slot sinx microdisk resonator. Opt. Express **18**(20), 20638–20644 (2010)
41. M.G. Scullion, T.F. Krauss, A. Di Falco, Slotted photonic crystals for biosensing applications. Proc. SPIE **8425**, 842507–842508 (2012)
42. C.A. Barrios, K.B. Gylfason, B. Sanchez, A. Griol, H. Sohlstrom, M. Holgado, R. Casquel, Slot-waveguide biochemical sensor. Opt. Lett. **32**(21), 3080–3082 (2007)
43. K. De Vos, I. Bartolozzi, E. Schacht, P. Bienstman, R. Baets, Silicon-on-insulator microring resonator for sensitive and label-free biosensing. Opt. Express **15**(12), 7610–7615 (2007)
44. K. Woo, K.J. Sheng Kee, Y. Shin, K. Han, Y. Yoon, G. Qiang, L.Q. Liu, X. Tu, M. Kyoung Park, Highly sensitive machzehnder interferometer biosensor based on silicon nitride slot waveguide. Sens. Actuators B: Chem. **188**, 681–688 (2013)

45. A. B. Gonzalez-Guerrero J. Osmond K. Zinoviev B. Sepulveda C. Dominguez S. Dante, D. Duval, L. M. Lechuga, Towards a complete lab-on-chip system using integrated mach-zehnder interferometers. Opt Pure Appl. **45**:87–95, 2012
46. S. Xiao, N. Mortensen, Highly dispersive photonic band-gap-edge optofluidic biosensors. J. Eur. Opt. Soc. Rapid publications **1**(0) (2006)
47. C. Smith, M.W. Lee, S. Tomljenovic-Hanic, C. Grillet, C. Monat, L. O'Faolain, C. Karnutsch, T.F. Krauss, R.C. McPhedran, U. Bog, L. Cameron, B.J. Eggleton, High-q microfluidic cavities in silicon-based two-dimensional photonic crystal structures. Opt. Lett. **33**(19), 2206–2208 (2008)
48. T.-Y. Chang, M. Huang, A.A. Yanik, H. Altug, Sub-wavelength nanofluidics in photonic crystal sensors. Opt. Express **17**(26), 24224–24233 (2009)
49. E. Hallynck, P. Bienstman, Photonic crystal biosensor based on angular spectrumanalysis. **18**(17), 1816418170 (2010)
50. M.R. Lee, P.M. Fauchet, Two-dimensional silicon photonic crystal based biosensing platform for protein detection. Opt. Express **15**(8), 4530–4535 (2007)
51. M.A. Dündar, R. Nötzel, M.J. van der Hoek, S. He, B. Wang, T. Siahaan, R.W. van der Heijden, Photonic crystal cavity on optical fiber facet for refractive index sensing. Opt. Lett. **37**(5), 833–835 (2012)
52. K.L. Watkin, L.L. Chan, S.L. Gosangari, B.T. Cunningham, A label-free photonic crystal biosensor imaging method for detection of cancer cell cytotoxicity and proliferation. Apoptosis **12**(6), 1061–1068 (2007)
53. D. Stellinga, M.G. Scullion, G.J.O. Evans, G.J. Triggs, M. Fischer, T.F. Krauss, Spatial resolution and refractive index contrast of resonant photonic crystal surfaces for biosensing. IEEE Photon. J. **7**(3), 1–10 (2015)
54. W.C Lai, C.Y. Lin, Y. Zou, S. Chakravarty, R.T. Chen, Methods to array photonic crystal microcavities for high throughput high sensitivity biosensing on a silicon-chip based platform. Lab Chip **12**, 2309–2312 (2012)
55. Genalyte. Genalyte, 2009, http://www.http://genalyte.com/
56. W. Bogaerts, P. De Heyn, T. Van Vaerenbergh, K. De Vos, S. Kumar, T. Selvaraja, P. Claes, P.Bienstman Dumon, D. Van Thourhout, R. Baets, Silicon microring resonators. Laser Photon. Rev.**6**(1), 47–73 (2012)
57. K. De Vos, Label-free silicon photonics biosensor platform with microring resonators. PhD thesis, University of Ghent, 2010
58. T. Klaes, Advanced silicon photonic ring resonator label-free biosensors. *University of Ghent*, 2012
59. A. Densemore, D.X. Xu, P. Waldron, S. Janz, P. Cheben, J. Lapointe, A. Delage, B. Lamontagne, J.H. Schmid, E. Post, A silicon-on-insulator photonic wire based evanescent field sensor. IEEE Photon. Technol. Lett. **18**, 2520–2522 (2006)
60. J.C. Aldridge, T.A Desai, J. Hryniewicz, N. Chbouki, B.E. Little, O. King V. Van, S. Chu, D. Gill, M. Anthes-Washburn, M.S. Unlu, A. Yalcin, K.C. Popat, B.B. Goldberg, Optical sensing of biomolecules using microring resonators. IEEE J. Sel. Top. Quant. Electron. **12**(1), 148–155 (2006)
61. B. Spaugh, F. Tybor, W.G. Gunn, M. Hochberg, T. Baehr-Jones, R.C. Bailey, M. Iqbal, M.A. Gleeson, L.C. Gunn, Label-free biosensor arrays based on silicon ring resonators and high-speed optical scanning instrumentation. IEEE J. Sel. Topics Quantum Electron. **16**(3), 654–661 (2010)
62. S. Choi, I. Jang, J.S. Choi, H.I Jung, H.J. Lee, J.H. Lee, Asymmetric split-ring resonator-based biosensor for detection of label-free stress biomarkers. Appl. Phys. Lett. *103*(5) (2013)
63. A. Kazmierczak, F. Dortu, M.J. Banuls, Polo A. Maquieira Catala, G.M. Kresbach, H. Sohlstroem, T. Moh, L. Vivien, J. Popplewell, G. Ronan, C.A. Barrios, G. Stemmea, C.F. Carlborg, K.B. Gylfason, W. van der Wijngaarta, A packaged optical slot-waveguide ring resonator sensor array for multiplex label-free assays in labs-on-chips. Lab Chip **10**(5), 281–290 (2010)

64. P. Taranekar, R.R. Ponnapati, W. Knoll, A. Baba, R.C. Advincula, Electrochemical surface plasmon resonance (ec-spr) and waveguide enhanced glucose biosensing with n-alkylaminated polypyrrole/glucose oxidase multilayers. ACS Appl. Mater. Interfaces **2**(8), 2347–2354 (2010)
65. Michael Rodahl, Bengt Kasemo, A simple setup to simultaneously measure the resonant frequency and the absolute dissipation factor of a quartz crystal microbalance. Rev. Sci. Instrum. **67**(9), 3238–3241 (1996)
66. Yong Chen, Hai Ming, Review of surface plasmon resonance and localized surface plasmon resonance sensor. Photon. Sensors **2**(1), 37–49 (2012)
67. Long range surface plasmon resonance for increased sensitivity in living cell biosensing through greater probing depth. Sens. Actuators B: Chem. **174**, 94–101 (2012)
68. A. Reisinger, Characteristics of optical guided modes in lossy waveguides. Appl. Opt. **12**(5), 1015–1025 (1973)
69. S. Ehsan, Salamifar and Rebecca Y. Lai. Application of electrochemical surface plasmon resonance spectroscopy for characterization of electrochemical DNA sensors. Colloids Surfaces B: Biointerfaces **122**, 835–839 (2014)
70. L. Coche-Guérentee, S. Armand, P. Labbé, A. Bouchet-Spinelli, B. Reuillard, S. Fort, Oligosaccharide biosensor for direct monitoring of enzymatic activities using qcm-d. Biosens. Bioelectron. **49**, 290–296 (2013)
71. J.A. Hubbell J.P. Bearinger, J. Vrs, M. Textor, Electrochemical optical waveguide lightmode spectroscopy (ec-owls): A pilot study using evanescent-field optical sensing under voltage control to monitor polycationic polymer adsorption onto indium tin oxide (ito)-coated waveguide chips. Biotechnol. Bioeng. **82**(4), 465–473 (2003)
72. J.J. Ramsden, Review of new experimental techniques for investigating random sequential adsorption. J. Stat. Phys. **73**(5), 853–877 (1993a)
73. A. Marti, M. Textor, P. Tengvall, J.J. Ramsden, R. Kurrat, B. Wlivaara, N.D. Spencer, Plasma protein adsorption on titanium: comparative in situ studies using optical waveguide lightmode spectroscopy and ellipsometry. Colloids Surfaces B: Biointerfaces **11**(4), 187–201 (1998)
74. A.D. Weston, L. Hood, Systems biology, proteomics, and the future of health care: toward predictive, preventative, and personalized medicine. J. Proteome Res. **2**(3), 179–196 (2004)
75. A.D. Edwards, Steric hindrance effects on surface reactions: applications to biacore. J. Math. Biol. **55**(4), 517–539 (2007)
76. R.A. Edwards, R.E. Huber, Surface denaturation of proteins: the thermal inactivation of beta-galactosidase (escherichia coli) on wall-liquid surfaces. Biochem Cell Biol. **70**(1), 63–69 (1992)
77. P. Colpo, A. Ruiz, L. Ceriotti, F. Rossi, *Surface Functionalization for Protein and Cell Patterning* (Springer, Berlin, 2010), pp. 109–130
78. A. Ulman, Formation and structure of self-assembled monolayers. Chem. Rev. **96**, 1533–1554 (1996)
79. W.C. Bigelow, D.L. Pickett, W.A. Zisman, Oleophobic monolayers. J. Colloid Sci. **1**(6), 513–538 (1946)
80. Seokheun Choi, Junseok Chae, Methods of reducing non-specific adsorption in microfluidic biosensors. J. Micromech. Microeng. **20**(7), 075015 (2010)
81. Y. Tai Tao, S.R. Wasserman, G.M. Whitesides, Structure and reactivity of alkylsiloxane monolayers formed by reaction of alkyltrichlorosilanes on silicon substrates. Langmuir **5**(4), 1074–1087 (1989)
82. T. Okubo, Y. Yamaguchi, M. Hu, S. Noda, Hiroshi Komiyama, Structure and morphology of self-assembled 3-mercaptopropyltrimethoxysilane layers on silicon oxide. Appl. Surface Sci. **181**(34), 307–316 (2001)

Chapter 3
Fabrication and Experimental Techniques

This chapter gives an overview of the fundamentals of the nanofabrication and experimental techniques employed throughout this thesis. The equipment used to realise the biosensors is presented along with details of the optimised fabrication processes. The chapter also discusses the fundamentals of electrochemistry, with an emphasis on semiconducting electrodes, together with the techniques used to characterise the electrochemical properties of the electro-photonic devices. Finally, the experimental setup which enables characterisation of the dual-mode optical/electrochemical devices is presented.

3.1 Introduction

Device fabrication was performed in the clean room (class 10,000) facilities of the Nanocentre at the University of York, whereas the optical and electrochemical characterisation was performed in the photonics (Physics) and the bio-inspired technologies (Electronics) laboratories, respectively.

Here we introduce the basics of each of the techniques used to fabricate our devices, providing references in case further detail is desired. Further details about our devices are provided in Chap. 4.

3.2 Nanofabrication Processes

In this section all the nanofabrication processes used to fabricate our sensor samples are presented. All of these devices were fabricated on a SOI material, which consists of a 220 nm top silicon layer formed on top of a 2 μm layer of silica (SiO_2) separating

J. Juan Colaś , *Dual-Mode Electro-photonic Silicon Biosensors*, Springer Theses,
DOI 10.1007/978-3-319-60501-2_3

it from a bulk silicon substrate. The order of the following subsections follows the order of the optimised fabrication protocol.

3.2.1 Diffusion Doping

The modification of the electric properties of a semiconductor substrate (e.g. silicon) can be tuned by the introduction of certain impurities (from another material) into the intrinsic crystalline structure of the substrate. This process, called doping, has been widely used over the last 80 years by the electronic circuit manufacturing industry [1]. Doping can be performed in two ways, ion implantation or diffusion. On incorporation into the crystal lattice, the dopant either gives up (donor) or receives (acceptor) an electron from the crystal. Adding either electrons or holes varies the conductivity of the semiconductor substrate, in this way adding electrons makes the material n-type, whereas adding holes makes the material p-type. Such introduction of impurities has therefore an effect on the Fermi level of the semiconductor as impurities add extra levels (illustrated in Fig. 3.1).

As observed in Fig. 3.1, in *n*-type doping donor ions are energetically close to the bottom of the conduction band. Therefore, carriers (electrons in this case) can be easily excited from the doping levels into the conduction band, leading to a shift of the Fermi level. The opposite occurs for *p*-type, where acceptor ions are energetically close to the bottom of the valence band, thus holes can be easily excited into the valence band leaving mobile holes. This excitation of holes leads to a downwards shift of the Fermi level. Here, we are interested in *n*-type doping as it presents higher conductivity and electrons are the major charge carries, being thus more suitable for the electrochemical processes we will work with.

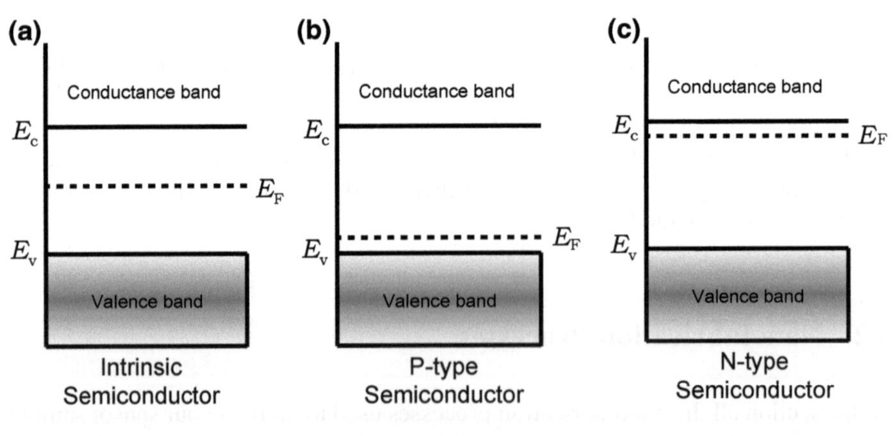

Fig. 3.1 Band structures for **a** intrinsic silicon, **b** *p*-type doped silicon and **c** *n*-type doped silicon. Extra levels are added by the impurities, which shift the Fermi level towards the conduction or valence band for *n*-type or *p*-type doping, respectively

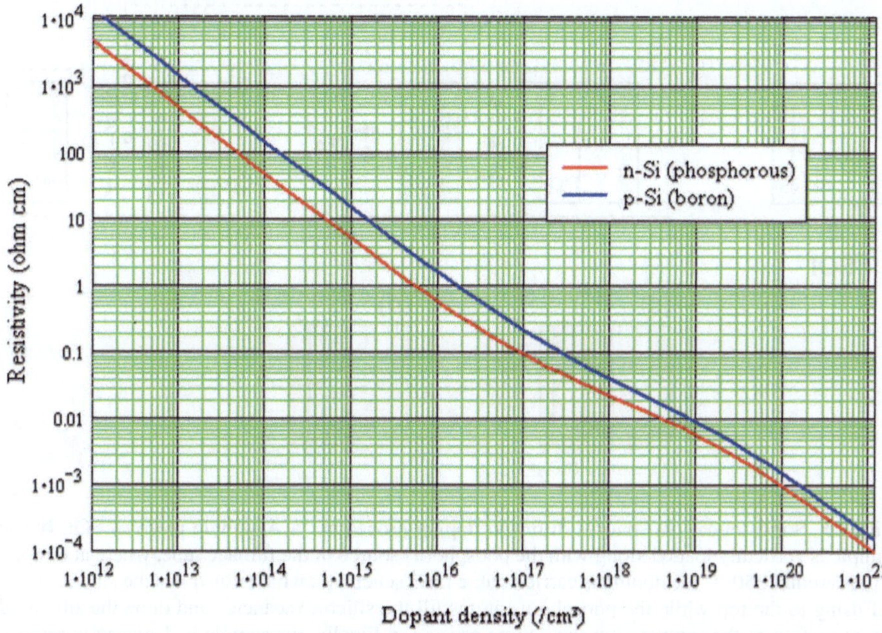

Fig. 3.2 Silicon resistivity as a function of doping concentration for both *n*-type (*red*) and *p*-type (*blue*) at 300 °K. Large doping concentrations present a low electrical resistivity, while low doping concentration show large electrical resistivity. Reproduced from pveducation.org

On the other hand, the quantity of dopant material introduced to a semiconductor substrate determines its concentration and indirectly affects many of its electrical properties. Depending on the level of doping (i.e. doping concentration, carrier concentration or impurities density), the electrical resistivity of the semiconductor material will have a larger or smaller value [2, 3]. For large doping concentrations, the substrate will present a low electrical resistivity; whereas, on the other hand, low doping concentration have large electrical resistivity (Fig. 3.2).

Figure 3.2 shows that for a given doping density, the electrical resistivity of a *n*-type doped silicon substrate is lower than that for *p*-type doping. As discussed in Sect. 4.2, the doping density of the electro-photonic sensor must be minimised in order to limit optical losses. For this reason *n*-type doped substrates were chosen.

In our fabrication process, these *n*-type impurities have been introduced through a diffusion method [4], where impurities are driven into the substrate from a solid source (i.e. dopant source). For *n*-type doping, phosphorous is used as the source. We use a solid source consisting of a ceramic wafer with the active ingredient Silicon Pyrophosphate (SiP_2O_7) carried on an inert silicon carbide (SiC) substrate (Phosphorus Grade PH-950, Saint-Gobain Ceramics, USA). The overall doping procedure is illustrated in Fig. 3.3.

As shown in Fig. 3.3, the SOI sample and the phosphorus source are mounted vertically in the furnace on a quartz boat with the phosphorus source placed upstream

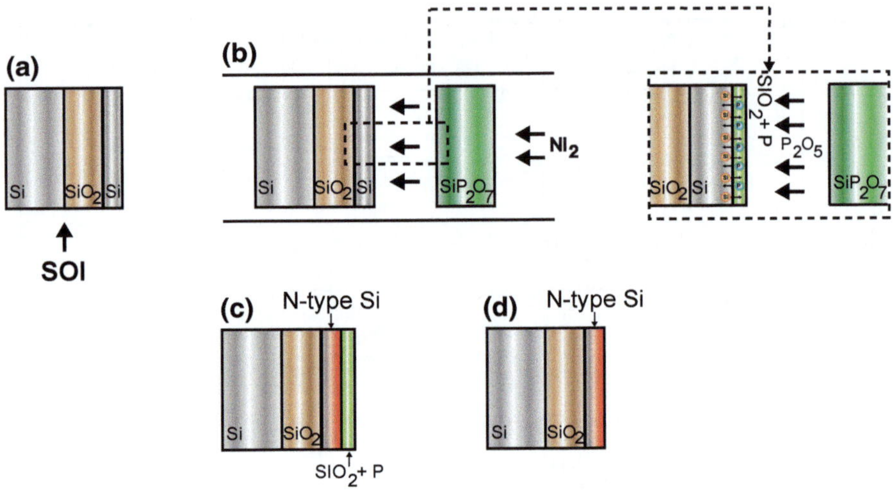

Fig. 3.3 Solid source phosphorus diffusion doping procedure. **a** A sample piece of SOI. **b** The sample is vertically loaded along with the phosphorus source in the furnace tube, where at temperatures around 850 °C the doping is carried out. **c** Inset schematic which illustrates the silicon atoms diffusing to the top while the phosphorus atoms fill the silicon vacancies and dope the silicon. **d** Cross section of the sample after the doping process. **d** Finally, the sample is deglazed to remove the oxide product through a dip in diluted HF (1:6 in DI water)

in the nitrogen flow. It is important that both sample and source are placed as close as possible without touching to ensure efficient diffusion of dopants to the substrate. At high temperatures, the active component of the silicon pyrophosphate decomposes to phosphorus pentoxide glass vapour (P_2O_5), which evaporates from the source by direct volatilisation and then condenses on the silicon wafer. This thin layer of P_2O_5 provides elemental phosphorus for the diffusion into the silicon via the reaction:

$$2P_2O_5 + 5Si = 4P + 5SiO_2 \tag{3.1}$$

At high temperatures, silicon atoms at the surface form silicon dioxide (SiO_2), leaving behind silicon vacancies. The phosphorous atoms then diffuse into the silicon layer and occupy the vacant sites left by the silicon atoms.

The process starts by pre-conditioning the dopant source. Sample and source are slid into the centre of the furnace tube where the temperature is best controlled. The temperature is slowly ramped at 5 °C/min from room temperature to the doping temperature (845 °C) in order to dry out any water molecules present in the source. The entire process is carried out in a 100% nitrogen environment, with a gas flow rate of 0.2 slpm (standard litre per minute). This slow ramping of the temperature is critical to ensure reproducibility between fabrication batches of −/+ 2% (when comparing the sheet resistance of the doped sample). Specific details of the thermal diffusion process are give in Sect. 4.2.

3.2.2 Thermal Evaporation

After modifying the electrical properties of the silicon substrate, electrical contact pads need to be fabricated on the doped silicon to enable electrons to be efficiently driven into and out of the substrate. These pads are deposited on the sample by means of resistive thermal evaporation.

In thermal evaporation, which is one of the most common and simple metal deposition processes [5], evaporation of the source metal is achieved by passing a large current through a boat or coil made of tungsten or similar material, in which the material to be evaporated is placed. When a certain temperature is reached (typically above 600 °C) at low chamber pressure, the material in the boat vaporises and re-condenses on the cooler sample which is placed overhead. This technique shows excellent performance for metals with relatively low melting points such as aluminium (660 °C), nickel (1455 °C) and gold (1064 °C).

The process was carried out using a Mantis HEX resistive thermal evaporator, which is fitted with three thermal boat sources allowing multi-layer stacks to be formed on top of the loaded substrates without the formation of intermediate native oxide layers. The chamber is pumped down to a pressure of order 10^{-6} mbar before the boats are heated, wherein previously high purity pieces of metals had been loaded as detailed in Sect. 4.3. The deposition rate and the film thickness is monitored in real-time by a quartz crystal microbalance (QCM).

Similarly as in the doping procedure, pre-conditioning is required to ensure good quality films. Both pressure and deposition rate are monitored while the current is increased in small steps (Fig. 3.4).

As illustrated in Fig. 3.4, at a certain current value, the pressure increases due the degassing of the metal piece. After the pressure reaches a local maximum value, it subsequently decreases and the deposition rate is triggered. It is at this point when the shutter of the chamber is opened and the vaporised metal is deposited on the substrate.

3.2.3 Electron Beam Lithography

By definition, lithography is a technique used for transferring a pattern into a material. In this thesis, the main lithography technique is electron beam lithography (EBL). For this process, a substrate is coated with a thin layer of electron beam (E-beam) sensitive resist. When this resist is exposed to an E-Beam, its chemical properties are changed. Depending on the type of resist, the exposed (non-exposed) areas can be dissolved in the proper solvent for a positive (negative) resist. For a positive resist, electrons break molecular bonds in the resist polymer to increase solubility. For a negative resist, the polymer molecules are cross-linked by electron exposure, transforming them into an insoluble material. As in photographic films, the step of removing exposed regions from the resist film with a solvent is called development. Finally, after development,

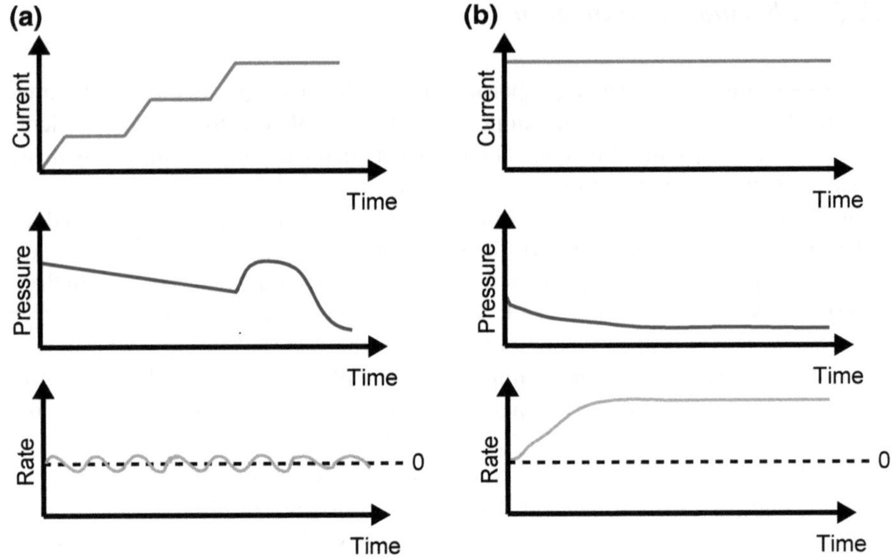

Fig. 3.4 Monitoring of evaporation process. **a** The current is ramped up until the pressure suddenly increases due to the degassing of the metal piece. **b** Once the pressure decreases, the evaporation is triggered

the remaining resist forms a pattern, which can be transferred into the underlying substrate via either dry or wet etching.

The resist used here is AR-P 6200.09 (ALLRESIST) [6], which is positive tone (exposed regions are removed) with high sensitivity to electron exposure, good resistance to plasma etching, and enable patterning of small features down to 6 nm. The resist layer is applied to the substrates using a spin-coater to ensure very flat and uniform resist layer of a specific thickness.

EBL was used to define all the photonic patterns on top of the silicon substrates. In EBL, a tightly-focused beam of electrons is accelerated towards the sample to expose particular areas of the resist. The beam of electrons is controlled and guided across the resist using electromagnetic lenses and electrostatic defection plates. Due to the short wavelength of high-energy electrons, this electron beam can be focused down to a few nm allowing the exposure of patterns with high spatial resolution. Nevertheless, and in contrast with optical systems, the smallest feature which can be exposed is limited by the diffusion and scattering of the electrons within the resist [5]. In particular, electrons that are back-scattered from the substrate result not only in increase exposure at the point of incidence of the electron beam, but they also expose areas that are some distance away, leading to a broadening of the exposed area. This effect (i.e. the additional exposure) is known as the electron beam proximity effect.

This proximity effect ultimately leads to areas in the lithographed pattern with incorrect dimensions. In order to obtain the desired dimensions in our structures, we manually adjusted the dose in some areas of our patterns (proximity error correction).

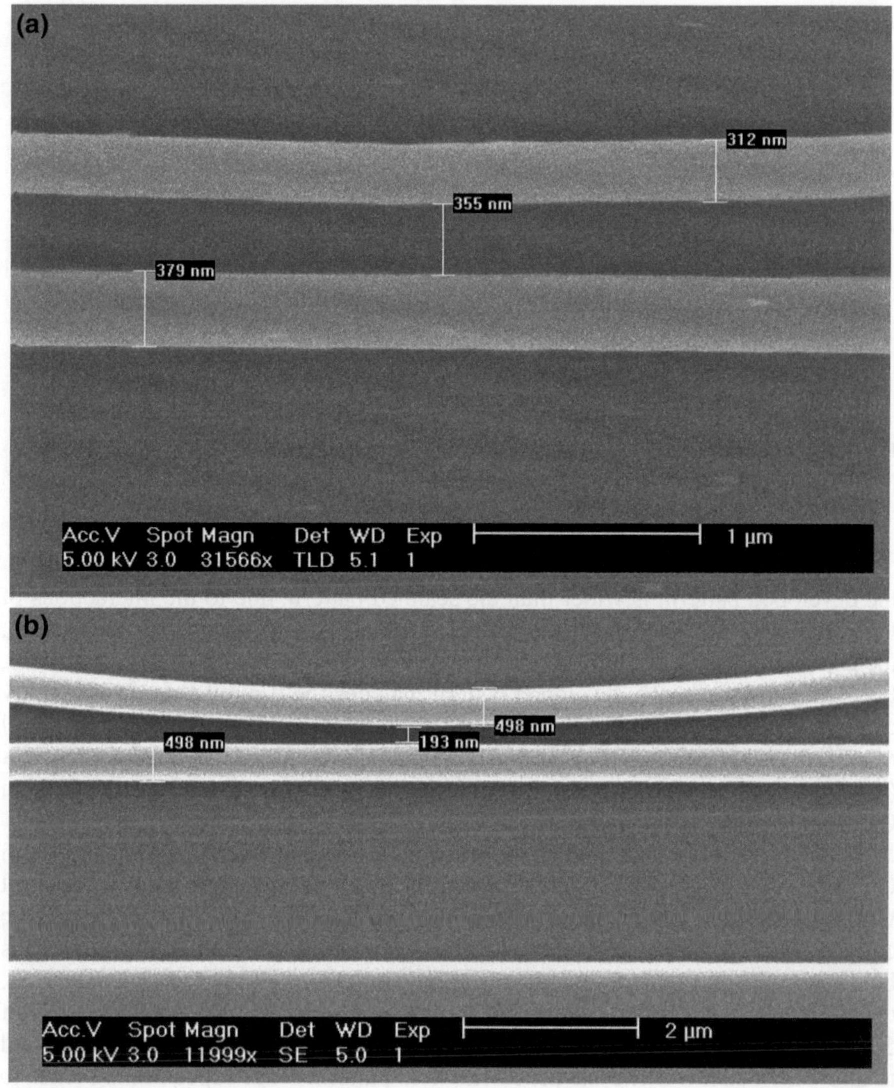

Fig. 3.5 Electron beam proximity effect in the coupling section of a ring resonator. **a** Backscattered electrons from large exposed areas that are at a short distance away from each other widen the exposed area. **b** After manually adjusting the exposure dose of these large exposed areas the desired structure dimensions are obtained

Figure 3.5 shows two images before (Fig. 3.5a) and after (Fig. 3.5b) adjusting the doses to compensate for the proximity effect.

In Fig. 3.5a, a pattern of a coupling section of a ring resonator with waveguide width of 500 nm and gap distance of 200 nm between waveguides is exposed. Duc to the proximity between the exposed areas, the exposure dose distribution, and hence

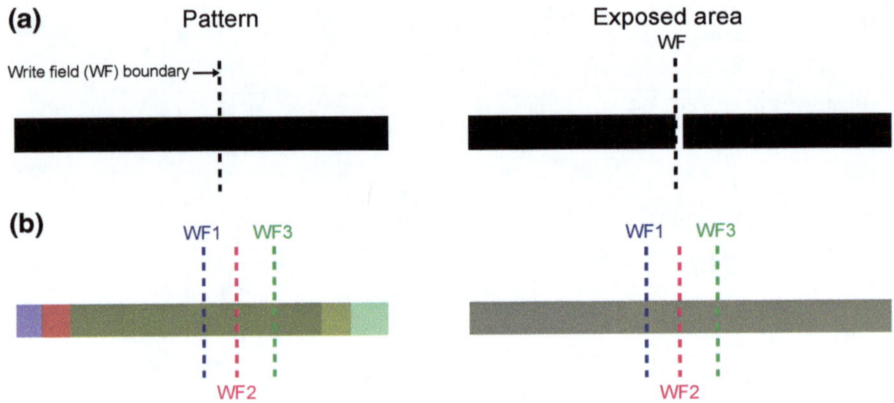

Fig. 3.6 Shot shifting technique to reduce stitching errors between write fields. **a** Alignment mismatches between write fields can lead to stitching errors. **b** Through a short-shifting technique, these errors can be minimised to decrease scattering losses

the developed pattern, is wider than the scanned pattern, due to the interactions of the primary beam electrons with the resist and substrate. To correct the pattern, the exposure doses were adjusted to obtained the desired pattern. After correction, Fig. 3.5b shows how the developed pattern matches the desired pattern. Proximity correction allows to control the geometry of our devices, providing an extra level of control of the coupling gap. It thus allows the optimisation of the coupling loss (loss between access waveguide and the ring cavity), which results crucial in high Q-factor devices [7].

The EBL system employed in this work is a Voyager from the company Raith GmbH, which has an acceleration voltage of 50 kV. This system allows both fast writing speeds and large beam currents. The beam is steered by a fully electrostatic column within a maximum write field size of 500×500 μm. When a larger area is being exposed, the computer divides the area up into multiple write fields. In our case, the full pattern is 12 mm × 100 μm, and is thus exposed in 24 different write fields of 500 μm in size. After writing the first field, the stage and the sample are repositioned to allow the beam to write the next write field pattern. Stage repositioning leads to errors in the pattern at the write field boundaries, known as stitching errors. For the in-plane applications like waveguides, stitching errors often lead to large scattering losses of confined light. To reduce these stitching errors, the shot shifting technique was applied [8]. This technique consists of splitting the pattern into a group of X shortly shifted sub-patterns with a dose equal to $\frac{1}{X}$ of the overall exposure dose. In this way, the sub-patterns overlay constructively at the write field boundaries, thus the stitching error is minimised (Fig. 3.6).

We employed this shot shifting in our exposures, using three sub-patterns shifted 10 μm away from each other. This level of sub-division afforded complete elimination of the stitching errors as confirmed by optical examination. Finally, after successful

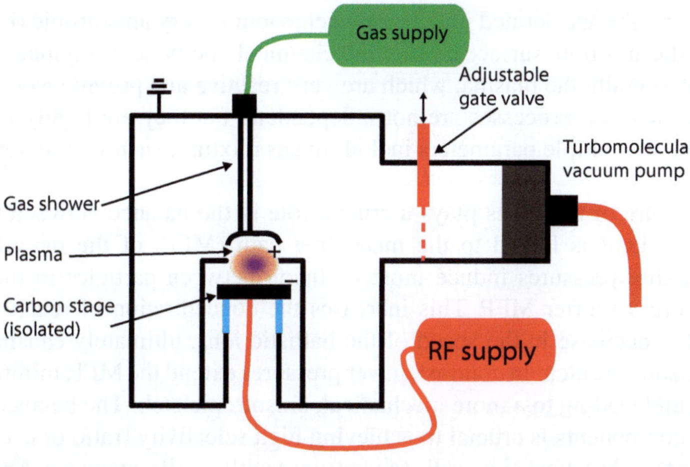

Fig. 3.7 Diagram of the in-house developed Reactive Ion Etching (RIE) system. Plasma is formed in the chamber to generate reactive ions and radicals which are bombarded into the sample to etch it. The pressure in the chamber is set using the gate valve

exposure, the samples are developed in Xylene to remove the resist from the exposed areas.

3.2.4 Reactive Ion Etching

Once a pattern has been lithographed on the silicon substrate, it needs to be transferred into the underlying substrate to create the desired structures. We employed a dry etching process widely used in the CMOS industry known as reactive ion etching (RIE) [9]. The RIE system is effectively a parallel plate capacitor in a vacuum chamber. The two plates of the capacitor consists of a metal disk with small holes where the reactive gases are fed into the chamber, while the bottom plate is a solid electrode where the sample is mounted (Fig. 3.7).

In the RIE system used for this work, which was developed in-house, the top plate (anode) is grounded, while the bottom plate (cathode) is electrically isolated from the rest of the system and coupled to an AC power supply which is driven at RF frequency (13.56 MHz). Under the application of a strong RF electromagnetic field, the gas between the electrode plates is decomposed and ionised, producing a plasma. This plasma generates reactive ions and radicals, such as fluorine (F^+), which are accelerated towards the sample surface. A characteristic glow can be observed due to the relaxation of gas molecules that have been excited inelastically by free electrons, with the emission colour being specific to the type of gas.

The RIE etching process has two components, the first being a mechanical sputtering process and the second the chemical etch component. The physical component is provided by the accelerated ions, whose behaviour is very anisotropic (i.e. etching normal to the substrate surface), while the chemical component originates from the gases used to ignite the plasma, which are very reactive and provide isotropic etching. These two etch processes are not independent, but they are highly interlinked and depend on multiple parameters including gas mixture, vacuum pressure and RF power.

The pressure of the gases plays a crucial role in the balance between these two components, as it is linked to the mean free path (MFP) of the particles in the plasma. Higher pressures induce more collision between particles in the plasma, and therefore a shorter MFP. This increases the concentration of reactive species along with a decrease in the speed of the ballistic ions, ultimately causing a more chemical, isotropic etch. In contrast, lower pressures extend the MFP, minimising ion scattering and leading to a more mechanical, anisotropic etch. The balance between these two components is crucial to achieving high selectivity (ratio of each rates of Si and resist) and vertical side walls (anisotropy) with small roughness. An excess of chemical etching will result in a loss of directionality with under-cut and angled side walls. On the other hand, a dominant physical component leads to angled side walls but with re-deposition of the etched material [9]. For this reason, a balance between both components needs to be satisfied to obtain vertical and smooth side walls.

Here, an etch process for silicon substrates was developed using CHF_3 and SF_6, as the source of reactive fluorine ions, while the H^+ ions contribute to polymer passivation and protect the material form further etching, ensuring smooth vertical side walls. The impact of the gas mixture, pressure and RF power were systematically explored until the optimal values for making optical devices in SOI were identified. In Fig. 3.8, a comparison between a pre-optimised and the optimised etch process is given for a grating pattern on a silicon substrate.

Following the optimisation process, the recipe consisted of a ratio of 1.16:1 for CHF_3 and SF_6 (respectively) setting a pressure of 4.2×10^{-2} mBar and a RF power of 22 W. Under these conditions, a DC bias of 188 V and an etch rate of 115 nm/min were obtained. I note that the roughness of the side walls could not be diminished adjusting the parameters previously described. This is due to the operation error of the gas mass flow controllers, which was around 10% of the defined gas flow value.

Once the sample is successfully etched, the remaining resist needs to be removed. This is achieved through the immersion of the sample in a solvent (1165 Microposit Remover) for 5 min under gentle sonication followed by immersion in acetone and isopropanol. Finally the sample is cleaved (using a diamond tipped scribe pen) in order to create the edge facets that allow the in-coupling of light.

3.2.5 Microfluidics

To ensure the delivery of the liquid sample to the sensor surface, a fluidic compartment was used to guide the fluids over the SOI samples. This compartment contains a fluid inlet and outlet port and a single broad channel through which the sample flows

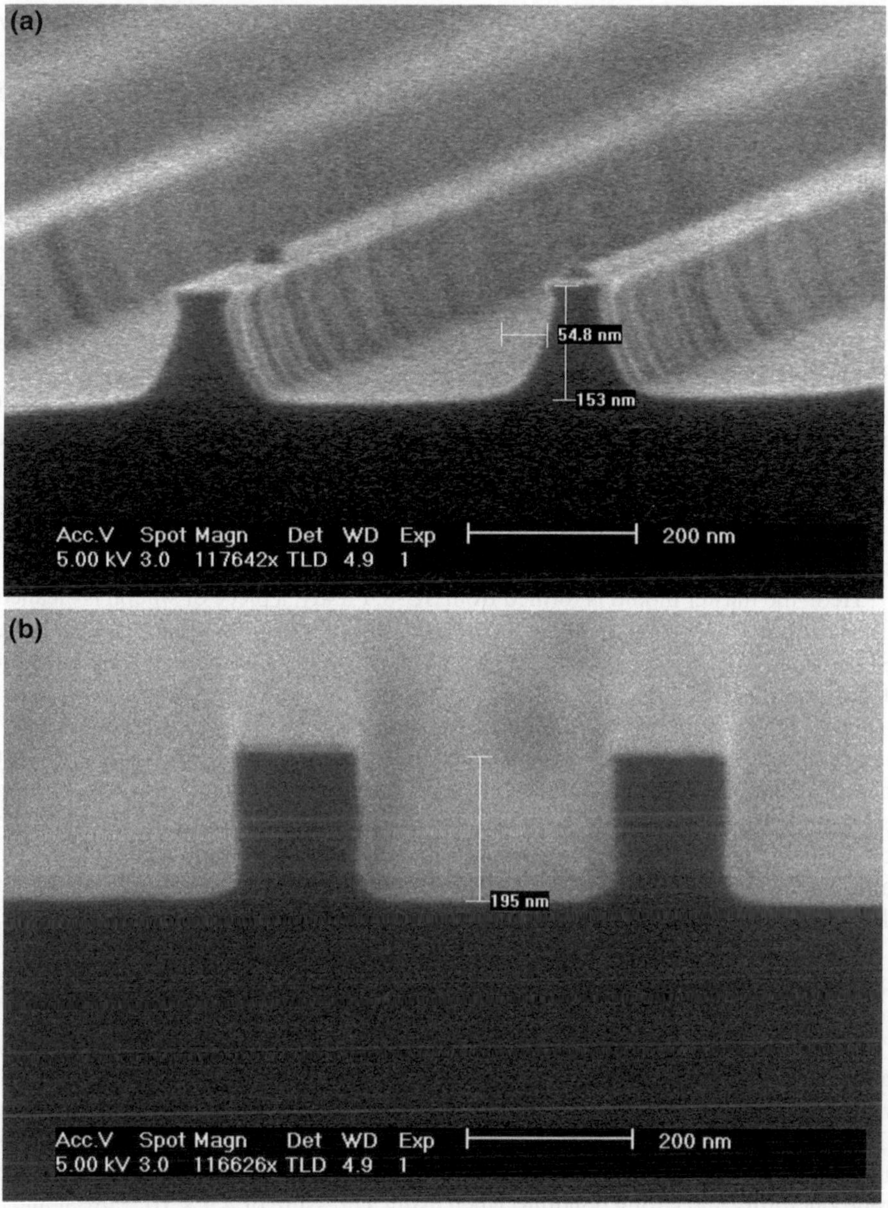

Fig. 3.8 Development of a silicon etch process using our in-house developed RIE. Along the development process, we observed how the excess of the chemical component results in an isotropic behaviour (a), while the proper balance between the two components process produces smooth vertical side walls

and interacts with the sensor surface. In order to provide controlled flow velocity and laminar flow, the channel had dimensions on the order of a few tens of micrometers, hence the name 'microfluidics'. A review of the state-of-the-art microfluidics techniques can be found in [10].

Microfluidics have been implemented using many different techniques and in a number of materials. One of the most common materials is poly(dimethylsiloxane), an elastomeric polymer commonly known as PDMS. PDMS has been the material used in this thesis to incorporate the microfluidics in our SOI chips. PDMS is optically transparent, flexible, gas permeable, and naturally hydrophobic, while being relatively cheap, simple and fast to use. Using a technique known as soft lithography, microfluidic devices can be rapidly fabricated in PDMS with feature sizes down to 30 nm [11].

PDMS (184 Sylgard) is made from two liquid components; a base elastomer and a curing agent. When these two parts are mixed and left to cure, a solid elastomer is formed. The curing process can be performed on top of a master mould which contains the complementary shape to the desired channel design, hence imprinting the shape of the mould in the cured PDMS when it is peeled away. This mould can be re-used multiple times to reliably make PDMS microchannels.

Here, the master mould is formed from SU-8. Firstly, a piece of silicon is cleaned in acetone and isopropanol under sonication and dried in nitrogen gas. Secondly, SU-8 negative photoresist (2050:2000.5 in a ratio of 10:1) is applied by spinning at 1000 rpm to obtain a thickness of 40 μm. Thirdly, the SU-8 coated silicon substrate is baked for 5 min at 65 °C, followed by 15 min at 95 °C. It is then exposed to UV light through a photomask of the channel pattern for 10 min. The sample is post baked for 1 min at 65 °C and 10 min at 95 °C. The pattern is obtained by developing the sample in EC solvent (Microposit, USA), a solvent engineered to eliminate the photoresist edge bead that occurs during typical spin coat wafer processing, for 5 min after which is rinsed in isopropanol. The sample is finally hard baked overnight at 180 °C, in order to ensure good adhesion of the SU-8 resist to the silicon substrate.

PDMS was obtained mixing the two agents in a ratio of 10:1.3 by volume (base and curing agent, respectively), and degassed in a desiccator for about 20 min to remove air bubbles. It was then poured over the master mould and left to cure for 5 h at 60 °C. Once solidified, access holes were punched at both ends of the microchannel and the PDMS piece was cut to size. To bond the PDMS block to the SOI chip, oxygen plasma was used for creating free OH groups on the PDMS surface (plasma activation) [12]. This was done using our in-house RIE system, where oxygen was used as the feed gas. Different recipes were explored, finding the best results (high reproducibility and strong bonding) when using a pressure of 9.9×10^{-2} mBar and a RF power of 22 W. The activation time was 20 s, as higher times lead to the formation of a glassy layer which decreases the quality of the adhesion [13]. After activation, the PDMS block is brought into contact and aligned with the SOI sample to from a covalent siloxane (Si-O-Si) irreversible bond [13]. Note that in parallel, the SOI sample has been piranha cleaned for 10 min (7:3 ratio H_2SO_4 to H_2O_2) to ensure that the surface is free from organic material, whilst adding OH groups on the silicon surface. Finally, it is placed in a 135 °C for 3 h to obtain a strong bond. Figure 3.9

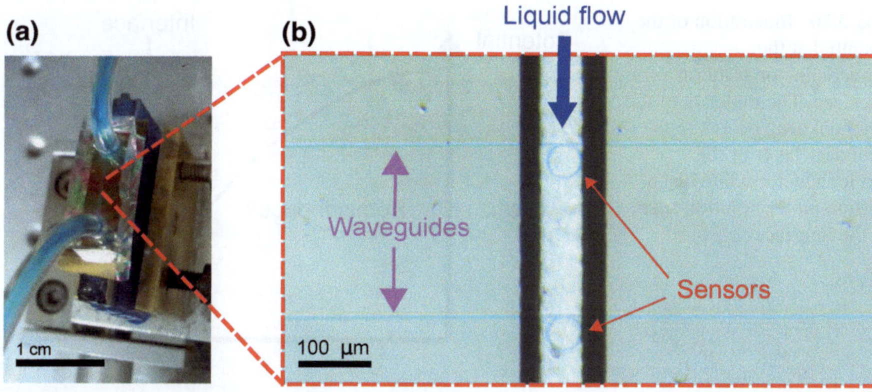

Fig. 3.9 SOI sample with integrated microfluidics. **a** A PDMS microchannel is bonded onto SOI to ensure the delivery of the liquid sample to the silicon sensor surface of the chip. **b** The microchannel is aligned to the optical structures which perform the sensing. Access waveguides take light to and from the sensors

shows two images of the finished sample. The diameter of the access holes is 2 mm, while an inlet tube diameter of 2.4 mm is employed to ensure the sealing of the channel. The volume of the flow cell is approximately 20 µL.

3.3 Electrochemistry

By definition, electrochemistry is the study of the interaction between electrical energy and chemical change. It is mostly focused on studying the principles of chemical reactions which take place at the interface of an electrode, typically a metal or a semiconductor, and an electrolyte [14]. Here, we provide a brief description of the processes involved at this interface, along with a presentation of the specific electrochemical technique that underpin this project.

3.3.1 Semiconductor-Electrolyte Interfaces

As at metal-electrolyte interfaces, when a semiconductor electrode is brought into contact with an electrolyte, electrons flow between these two media until equilibrium of the two electrochemical potentials is reached. The equilibrium is reached when the Fermi energies of the two media are equal, meaning that the electric field at all interior points must be zero (no net movement of charge carriers). The rearrangement of carriers occurs close to the interface, and directly depends on the particular mate-

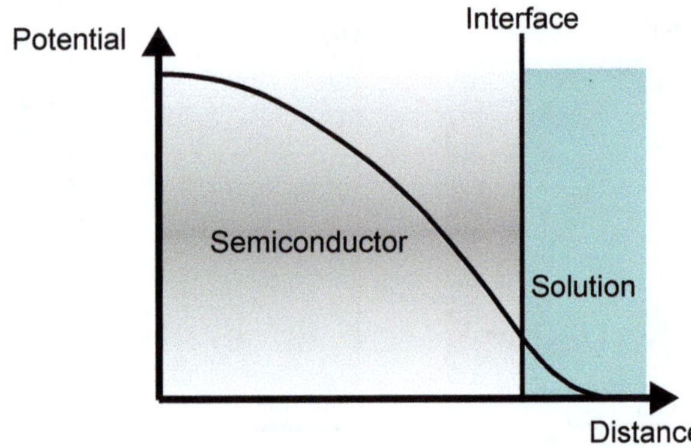

Fig. 3.10 Illustration of the potential at the semiconductor-solution interface. The majority of the potential drop occurs in the boundary layer of the semiconductor while just a portion on the solution side of the interface

rials (electrode and electrolyte) involved. This rearrangement results in a potential difference at the interface (interfacial potential).

Due to the fact that the conductivity of semiconductors is typically lower than that of most electrolytes, the majority of the potential drop occurs in the boundary layer of the semiconductor electrode while just a portion on the solution side of the interface [15] (Fig. 3.10). This is very similar to what occurs at semiconductor-metal interfaces Sect. 4.3.

This variation of the electrostatic potential ($\phi(x)$) around the interface results in a bending of the energy bands inside the semiconductor, as the potential contributes to the electronic energy ($-e_0\phi(x)$). The application of a potential difference (ϕ) across the semiconductor-electrolyte interface changes the electrostatic potential at the interface (ϕ_s), depending on the magnitude and direction of the applied bias. In a n-type semiconductor, if $\phi_s > 0$, the conduction and valence bands in the semiconductor bend downwards thus the concentration of electrons in the conduction band is enhanced (Fig. 3.11a). This interfacial region of high electron density is known as the "enrichment layer". If $\phi_s < 0$, the bands bend upwards, creating a 'depletion layer' at the surface where the concentration of electrons is reduced (Fig. 3.11b). With large biases, the concentration of the holes in the valence band can be enhanced at the electrode boundary, forming an "inversion layer" (Fig. 3.11c). Finally, the particular potential at which the electrostatic potential is constant throughout the semiconductor $\phi(x) = 0$, is known as the "flat-band potential", which equals the potential of zero charge.

In some cases, a molecule either in the electrolyte itself or immobilised to the electrode surface, can undergo reduction-oxidation (redox) reactions. These types of reactions involve an exchange of electrons between two species, here the semiconductor electrode and the redox active molecule. In these situations, the equilibrium between the charge on the semiconductor surface, and the localised concentration of redox species at the boundary of the electrolyte with the semiconductor, is reached when the Fermi energy is aligned in between the energies of the reduced and oxidised species (Fig. 3.12a). As the conductivity of doped semiconductors is usually well below that of the electrolyte solution, the positions of the band edges in the

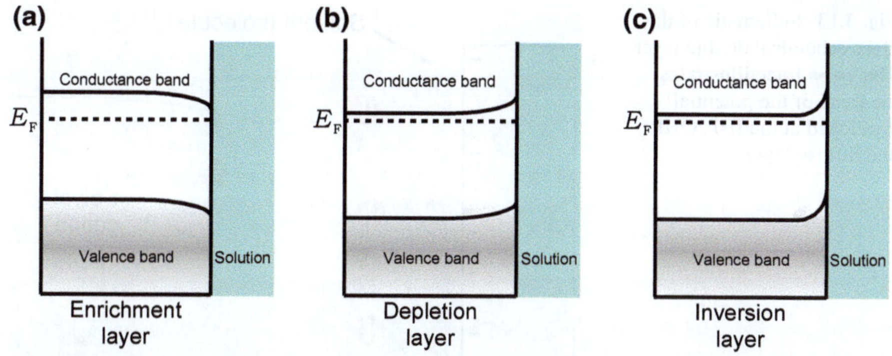

Fig. 3.11 Band bending at the interface between a *n*-type semiconductor and electrolyte. **a** Enrichment layer, **b** Depletion layer and **c** Inversion layer

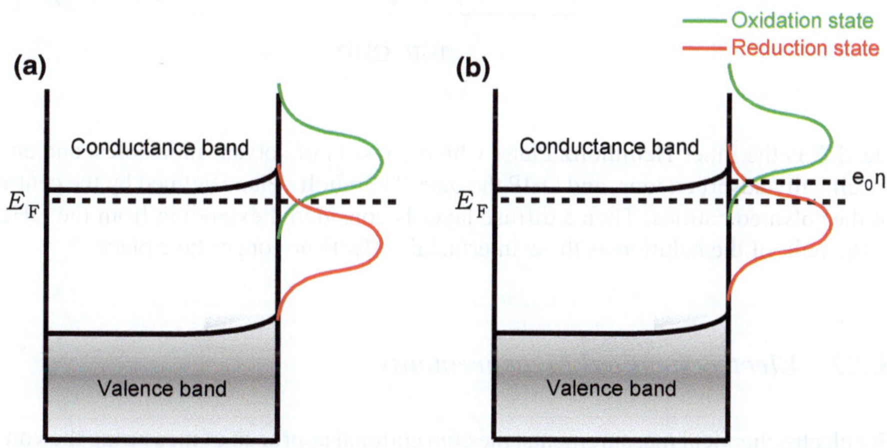

Fig. 3.12 Energy diagram for a redox reaction at the interface of an *n*-type semiconductor with an electrolyte. **a** At equilibrium, the Fermi levels of the semiconductor and of the redox species are equal. **b** After the application of a positive overpotential ($\eta > 0$), the relative position of the Fermi level in the semiconductor is changed

semiconductor region are not modified with respect to the electrolyte/immobilised molecule when the potential (η) is modified. However, the relative position of the Fermi level in the semiconductor is changed, and so is the density of carriers on the semiconductor surface (Fig. 3.12b).

The solution side of this interface also needs to be considered. When an electrode (either metal or semiconductor) is brought into electrical contact with a solution, an electrical double-layer capacitance will be formed on the electrode-electrolyte interface. The origin of this double-layer arises from the accumulation of solvated ions in the solution side to screen the electrode surface charge. The Helmholtz [16] model simulates this structure by a capacitor. Figure 3.13 displays its structure, where

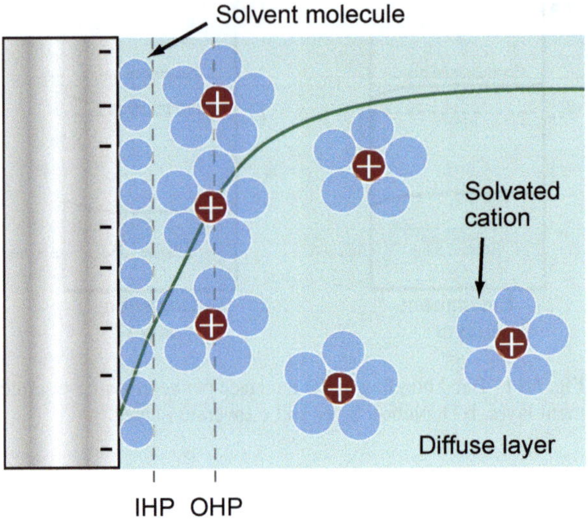

Fig. 3.13 Schematic of the electrochemical double layer. The *green* trace illustrates the trend of the potential developed at the IHP, OHP and diffuse layer

the IHP is the inner Helmholtz Plane which consists of solvent molecules, and any specifically adsorbed ions, and OHP the outer Helmholtz Plane defined by the centre of the solvated cations. Then a diffuse layer is considered extending from the OHP to the bulk of the solution as these interfacials effects no longer take place.

3.3.2 Electrochemical Measurements

For electrochemical measurements, the conventional approach of measuring the voltage drop across two points is invalid due to the potential that develops as a result of insertion of an electrode into an electrolyte. To account for this electrode potential, a third electrode needs to be included into the system. This electrode presents a potential relative to the solution which is known and stable, thus subsequent measurements will be normalised to it, and hence the name "reference electrode".

In most electrochemical measurements, three electrodes are employed in the measurement cell (Fig. 3.14). These electrodes are known as the working electrode (WE), the counter electrode (CE), and the reference electrode (RE). In this common arrangement, current is sourced from the counter electrode and reactions are measured at the working electrode (which is the silicon substrate in our work). The potential required at the working electrodes is supplied relative to the reference electrode.

Reference electrodes are commonly assembled with a metal and a metal salt surrounded by an electrolyte solution, and with a porous membrane allowing implementation of the ionic circuit [17]. Furthermore, they present a very high impedance to prevent current flow into them. In order to prevent leakage of soluble ions from the electrolyte of the reference electrode to the electrolyte in the three terminal electro-

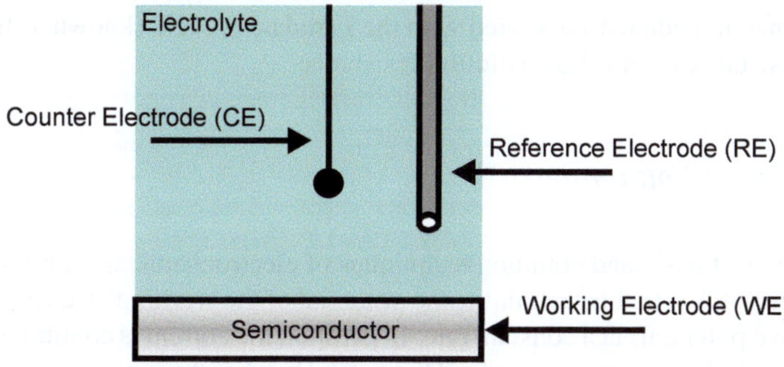

Fig. 3.14 Three terminal electrochemical cell. Reactions are measured at the working electrode (WE), and current is supplied at the counter electrode (CE). The potential required at the working electrodes is generated relative to the reference electrode (RE)

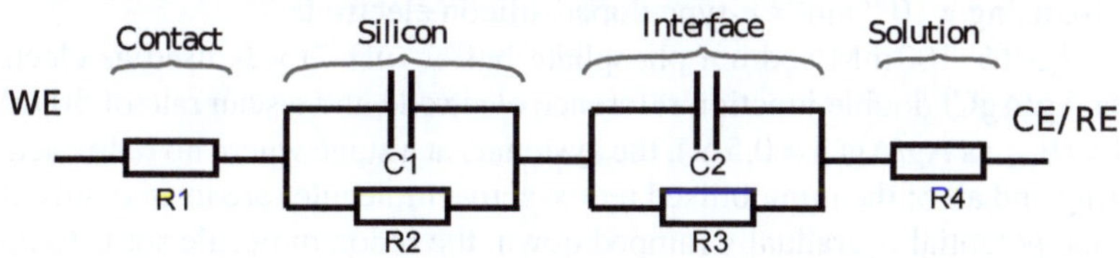

Fig. 3.15 Equivalent circuit based on the Randels' circuit adapted to semiconductor electrodes. R1 represents the Ohmic contact resistance and the resistance of the silicon, C1 and R2 represent the capacitance and the resistance of the depletion region of the silicon, respectively, C2 the double layer capacitance, R3 the resistance impedance associated with the Faradaic process and R4 the solution resistance

chemical cell, a double junction silver/silver chloride (Ag/AgCl) was employed in our setup. This double junction electrode consists of two electrodes, where one (inner electrode) is encapsulated in a glass shell inside the second one (outer electrode). In our case, both inner and outer electrodes contain the same electrolyte, which is KCl at a concentration of 3.5 M. Finally, the three electrodes are connected to a potentiostat (EmStat, Palmsens) which is able to measure current flow while applying a voltage relative to the reference electrode.

The complete three terminal setup can be decomposed into equivalent electrical elements that models the double layer capacitance, electrolyte resistance and the impedance associated with a redox reaction. For our system, we have employed equivalent circuit based on the commonly used Randels' circuit [18] adapted to model semiconductor electrodes (Fig. 3.15), which is provided to help understanding of the electrical properties of the system employed throughout this thesis. More complex electrical circuit models can be applied to precisely characterise the associated impedance/capacitance [19].

In this simple circuit model, R1 represents the Ohmic contact resistance and the resistance of the silicon, C1 and R2 represent the capacitance and the resistance of the depletion region of the silicon, respectively, C2 the double layer capacitance, R3 the resistance impedance associated with the Faradaic process (known as the charge transfer resistance) and R4 the solution resistance.

3.3.3 Cyclic Voltammetry

One of the most basic and common techniques of electrochemical analysis is Cyclic Voltammetry (CV). In this technique, the potential of the working electrode is cycled between two potentials at a constant rate. In parallel, the current is continuously monitored through the counter electrode. Through a Faradaic process, certain potentials at the working electrode may give electrons enough energy to tunnel in and out of the solution and change the redox state of the present species. An example of a CV measurement using a surface immobilised electroactive molecule [20] is shown in Fig. 3.16, using a 10^{20} cm^{-3} n-type doped silicon electrode.

In Fig. 3.16, 100 mM sodium phosphate buffer (pH 7) was used as electrolyte, using a Ag/AgCl double junction reference electrode and a scan rate of 50 mV s^{-1}. Initially (E_{we} vs Ag/AgCl = 0.5 V), the system is at a stage where no redox activity is occurring and all of the immobilised redox-active molecules are in an oxidised state. When the potential is gradually ramped down, the redox molecule starts to undergo reduction as electrons in the electrode begin to tunnel to the redox molecule. The rate of this reaction peaks, in Fig. 3.16, at a potential of E_{we} versus Ag/AgCl = −0.35 V. As the potential is decreased further, every single molecule on the surface is reduced, thus the redox current decreases as no further molecules can be reduced. When the direction of the potential sweep is reversed, the reverse reaction occurs; electrons now tunnel from the immobilised redox system to the electrode.

The separation between the peak oxidation and reduction potential relates to the over potential that needs to be applied so electrons can tunnel in and out of the silicon electrode through the depletion layer (for a metal electrode where there is no

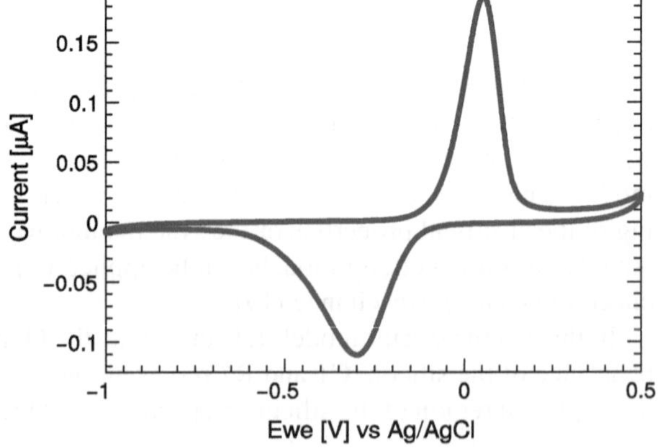

Fig. 3.16 Cyclic Voltammetry (CV) of the redox molecule published in [20] on a 10^{20} cm^{-3} n-type doped silicon substrate as electrode (100 mM sodium phosphate buffer (pH 7) as electrolyte) at a 50 mV s^{-1} scan rate. The positive and negative current peaks observed using are indicative of electrochemical oxidation and reduction of MB

surface depletion, the separation is ideally zero). The charge transferred during reduction/oxidation processes can be employed to calculate the total number of molecules immobilised on the surface as explained in Sect. 4.6. Cyclic voltammetry can also be used to determine different ions present in the solution (as different ions exhibit different redox potentials), and furthermore, it is a useful tool to assess the state of cleanliness of the working electrode.

3.4 Electro-Optical Characterisation Setup

Our experiments were carried out in a modified end fire (method of coupling two waveguides for optical fibers or integrated optical circuits by placing the two up against one another) transmission setup, modified to enable simultaneous electrochemical measurement. Light from a single mode fibre coupled to a broadband amplified spontaneous emission (ASE) source (1520–1620 nm, 54 mW peak optical power) was collected and collimated by an aspheric lens and splitter cube to a 60x lens. The spot from this lens was focused onto an access waveguide (approximately 5 μm width) on the edge facet of the fabricated SOI chip. The required alignment was performed using an infrared camera with a 20x objective, while illuminating the sample with a white light source. Light propagating through the sample to the back facet was then collected and collimated by a 40x lens, and sent through free space to a focusing aspheric lens onto the facet of a single mode fibre. This light was finally split using a splitter fibre to an optical spectrum analyser (OSA). Continuous measurements of the transmission spectrum were taken with the OSA, where the resonance wavelength is continuously Lorentzian fitted and monitored using Labview ™. A positive displacement syringe pump and sample injection valve (with 6 ports) were used to ensure a controlled delivery of samples through the inlet and outlet ports of the microfluidic channel. Unless otherwise stated, the flow rate employed throughout this thesis was 20 μL/min. The silicon substrate was used as the electrochemical working electrode and, via a potentiostat, the potential of the silicon electrode was controlled relative to that of an Ag/AgCl double junction reference electrode (with KCl at 3.5 M as electrolyte), while a platinum counter, electrode completed the circuit (Figs. 3.17 and 3.18). I note that no temperature control was incorporated in this setup as the high reproducibility of the results was deemed sufficient at this design/test phase. However, to further reduce the noise level of the system and therefore improve its performance in biosensing applications, a temperature control unit would probably be required.

3.5 Summary

In summary, the first half of this chapter is devoted to present the fundamentals of the fabrication process followed to fabricate our bi-domain sensors. Firstly, the silicon

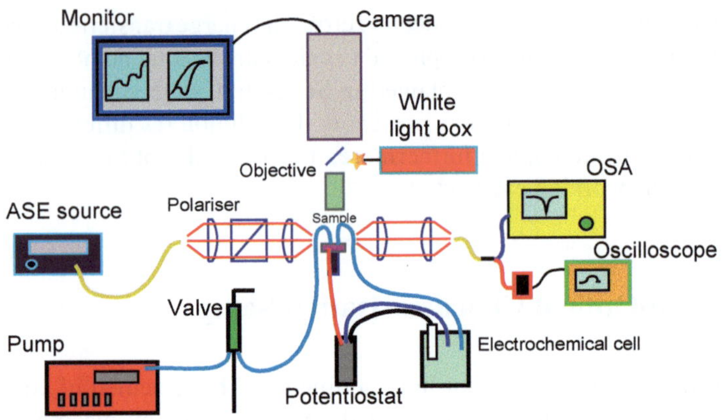

Fig. 3.17 Electro-optical characterisation setup. TE polarised light is focused onto sample edge facet using aspheric lenses. Light collected from back facet split between OSA and photodetector. The alignment is optimised using the photodetector signal. A syringe pump and a valve are used to control the fluidics, while the electrochemical functionalities are controlled via a potentiostat

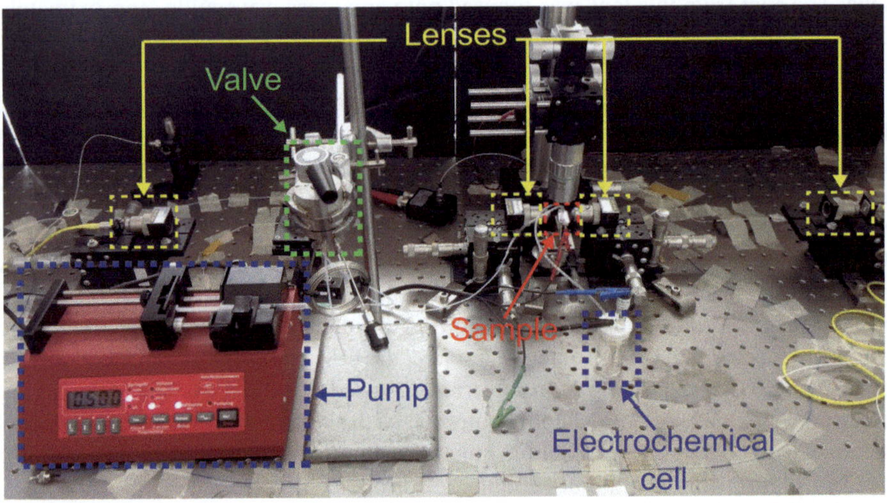

Fig. 3.18 Picture of the electro-optical characterisation setup. The main elements of the setup (lenses, microfluidics pump, valve, sample and electrochemical cell) have been highlighted

layer of the SOI substrate is doped to allow electrochemical measurements, which is then followed by a deposition of metal contacts thus electrons can efficiently tunnel into the substrate. Once the electrical properties are modified, optical structures are etched in this silicon layer trough electron beam lithography and reactive ion etching. Finally, to ensure adequate delivery of sample to the sensor surface, a PDMS microfluidic channel is bound to the device. The second half of the chapter describes the principles of electrochemistry, focusing on semiconductor electrodes. The electrical charge distribution on both sides of the semiconductor-electrolyte interface is presented, showing how the semiconductor side has a bigger impact on device performance. The requirements for electrochemical measurements are also discussed, and finally, the electrochemical technique used in our experiments (Cyclic Voltammetry) is introduced.

References

1. E. F. Schubert, *Doping in III-V Semiconductors* (Cambridge University Press, 1993)
2. Y.M. Liu, W.R. Thurber, R.L. Mattis, J.J. Filliben, Resistivitydopant density relationship for borondoped silicon. J. Electrochem. Soc. **127**, 2291–2294 (1980a)
3. Y.M. Liu, W.R. Thurber, R.L. Mattis, J.J. Filliben, Resistivity-dopant density relationship for phosphorus-doped silicon. J. Electrochem. Soc. **127**, 1807–1812 (1980b)
4. D. Mathiot, J.C. Pfister, Dopant diffusion in silicon: a consistent view involving nonequilibrium defects. J. Appl. Phys. **55**, 3518–3530 (1984)
5. M.J. Madou, *Fundamentals of Microfabrication and Nanotechnology*, 1st edn (CRC Press, 2011)
6. ALLRESIST. Positive E-Beam Resists AR-P 6200, year = 2016, http://www.allresist.com/wp-content/uploads/sites/2/2014/12/allresist.produktinfos.ar-p6200.englisch.pdf. Accessed 15 June 2016
7. Y.A. Vlasov, S. Assefa, C. Kang, C.T. Phare, S.M. Weiss, Photonic crystal slab sensor with enhanced surface area. Opt. Express **18**(26), 27930–27937 (2010a)
8. G. Pagnotta, L. OFaolain, D. OBrien, T.F. Krauss, Shot shifting for nanophotonic applications. Microelectron. Eng. **84**(58), 1463–1466 (2007)
9. F.F. Chen, *ntroduction to Plasma Physics and Controlled Fusion: Volume 1: Plasma Physics*, 2nd edn (Springer, 1984)
10. E.K. Sackmann, A.L. Fulton, D.J. Beebe, The present and future role of microfluidics in biomedical research. Naturc **507**, 181–189 (2014)
11. Y. Xia, G.M. Whitesides, Soft lithography. Annu. Rev. Mater. Sci. **28**, 153–184 (1998)
12. M.A. Johnson, M.A. Eddings, B.K. Gale, Determining the optimal pdms-pdms bonding technique for microfluidic devices. J. Micromech. Microeng. **18**, 067001 (2008)
13. B. Millare, M. Thomas, A. Ferreira, X. Hong, Madison Holesinger, Valentine I. Vullev, Dependence of the quality of adhesion between poly(dimethylsiloxane) and glass surfaces on the conditions of treatment with oxygen plasma. Langmuir **24**, 13218–13224 (2008)
14. A.J. Bard, L.R. Faulkner, *Electrochemical Methods, Fundamentals and Applications*, 1st edn. (Wiley, 2001)
15. K. Rajeshwar, *Fundamentals of Semiconductor Electrochemistry and Photoelectrochemistry*, 1st edn. (Wiley, 2007)
16. H. Helmholtz, Ueber einige gesetze der vertheilung elektrischer strme in krperlichen leitern mit anwendung auf die thierisch-elektrischen versuche. Annalen der Physik **165**(6), 211–233 (1853)
17. D.J.G. Ives, G.J. Janz, *Reference Electrodes, Theory and Practice*, 1st edn. (Academic Press, 1961)
18. J.E.B. Randles, Kinetics of rapid electrode reactions. Discussions of the Faraday Society **11**, (1947)
19. R. Hui, B. Wang, Q. Huang, J. Zhang, A review of ac impedance modeling. Electrochimica Acta **11**, 8145–8163 (2007)
20. L. Pukenas, R. Azhar, M. Guillorit, C. Walti, K. Critchley, S. Johnson, J. Murray, D. Nowak, R.S. Bon, Solid phase synthesis of functionalised sam-forming alkanethiol oligoethyleneglycols. J. Mater. Chem. B **2**, 3741–3744 (2014)

Chapter 4
The Electro-Photonic Silicon Biosensor

This chapter describes the characteristics of our dual-mode biosensor that combines electrochemical and photonic detection on the same silicon substrate. The chapter starts with a comparison between two different approaches for combining sensing domains and highlights their limitations and advantages. Following this comparison, the option followed in our work, namely selective doping of the silicon substrate, is studied and characterised in detail from electrical, electrochemical and photonic points of view. The optimised system is presented alongside a proof-of-principle experiment. The chapter concludes with a discussion of the limitations of our approach and a comparison of the electro-photonic sensor with related bi-domain sensing techniques.

4.1 Alternatives

Analytical technologies that combine information from multiple sensing domains offer a number of important advantages, as discussed in Sect. 2.4. For example, multi-domain technologies can provide deeper understanding of molecular and biomolecular processes [1–3]. In this section, we focus on technologies that enable the integration of electrochemical and photonic sensing techniques on the same silicon photonics platform.

The challenge associated with combing electrochemical and photonic sensing in a single platform concerns the appropriate choice of materials and sensor architecture. Photonic sensors require materials that can be patterned to create the desired optical structures and that exhibit a high refractive index contrast with the surrounding material to ensure confinement of the optical mode. In order to support electrochemical measurements, materials are required that facilitate electron transfer between the electrode and an analyte and are electrochemically inert over a wide potential window. Here I discuss two approaches for meeting these different material

J. Juan Colaś , *Dual-Mode Electro-photonic Silicon Biosensors*, Springer Theses,
DOI 10.1007/978-3-319-60501-2_4

requirements: (1) depositing an optically transparent and conductive layer on top of an optical compatible substrate, and (2), tuning the properties of an optical compatible substrate so that it also supports electron transport.

Conductive optically transparent layer on a silicon photonics device

Arguably the simplest approach for combining photonic and electrochemical sensing is to employ two materials, one to confine the optical mode and the other to facilitate electrochemical analysis, for example an SOI substrate coated with a layer of indium tin oxide (ITO). Here the SOI confines the optical signal for optical sensing while the electrically conductive and transparent ITO layer permits electrochemical measurement while enabling penetration of the optical mode into the supporting analyte. This approach is relatively simple to fabricate and is based on established materials. For example, ITO has been employed widely in the literature as an optically transparent electrode material [4], and has been demonstrated in a range of biosensing applications [5–7]. Furthermore, ITO is employed in Optical Waveguide Lightmode Spectroscopy (OWLS) in order to combine optical and electrochemical measurement [1], as discussed in Sect. 2.4.

In the SOI-ITO scheme, light coupled into the media will propagate through the patterned SOI substrate while electrochemical reactions will be controlled and measured at the ITO electrode. The evanescent tail of the optical signal will penetrate across the ITO layer sensing these reactions to transduce them to the optical domain. However, as a portion of the optical mode now propagates through the ITO film, there is potentially an associated reduction in sensitivity due to the reduced interaction between the optical mode and the analyte. To quantify this loss in sensitivity, 3D finite-difference time-domain (FDTD) simulations (using MEEP [8]) were performed to explore the viability of this approach.

In these simulations, a photonic crystal cavity device (in which a row of three holes is removed from the lattice, also known as "L3" type cavity) was chosen as the optical sensor [9]. This sensor was chosen for two reasons. Firstly, the L3 Phc device has a planar architecture ensuring that a continuous ITO conductive layer can be deposited across the sensor surface, facilitating electrical contact. Secondly, the optical mode is confined in the silicon rather than in air; in contrast, those PhC configurations where sensing is performed in an air cavity [10, 11] are incompatible with addition of a conductive layer.

Two different scenarios were considered to evaluate this technology. First, as shown in Fig. 4.1a, the sensitivity of the sensor to a uniform 15 nm thick layer of variable RI was evaluated. Here, the thickness of the overlayer was chosen to simulate a densely packed layer of antibodies immobilised on the sensor surface. In the second scenario (Fig. 4.1b), the sensitivity was assessed as a function of the thickness of a uniform overlayer of constant refractive index. Again, the RI of this layer was chosen to match that of a layer of antibodies. In both cases, a 30 nm thick film of ITO was considered as for such a thickness, the ITO electrical resistivity is sufficiently low [4] to support electrochemical reactions.

When the sensor with 30 nm thick film of ITO was simulated, a high reduction of the sensing performance was observed. A sensitivity of 2.95 nm/RIU was obtained,

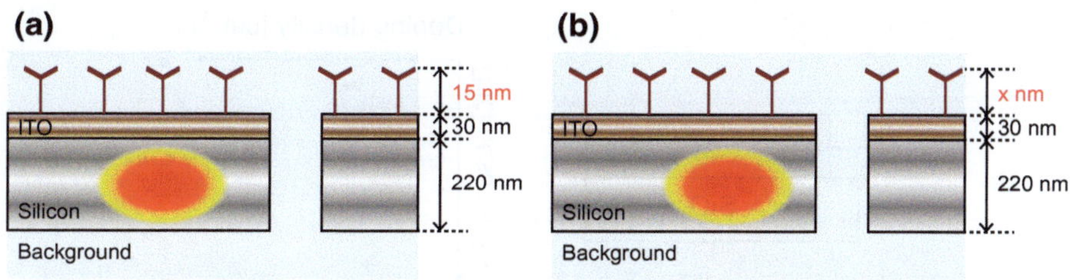

Fig. 4.1 Cross profile sections of the simulated L3 PhC biosensors scenarios

which compares poorly to the 5.46 nm/RIU observed with the SOI device without ITO film. In the second situation, 0.11 nm/nm (nm of optical shift against nm of overlayer thickness) was achieved, corresponding to a 35% reduction in sensitivity. These results clearly demonstrate that the sensing performance is highly diminished by the addition of the ITO layer and is thus not ideal for the realisation of a highly sensitive multi-modal sensor.

Single electrochemically and optically compatible substrate

As an alternative to employing two materials, one for each sensing domain, I investigated multi-modal sensing based on a single material whose properties have been tuned to support both photonic and electrochemical measurements. SOI is an excellent candidate material to meet this criterion. Silicon is a semiconductor with a refractive index around 3.47 (at 1550 nm). It is thus ideally suited for the fabrication of integrated optical components. Furthermore, by doping of the silicon substrate, it is possible to tune its conductivity sufficiently to enable its use as an electrochemical electrode [12]. This modification of the electrical properties of the substrate is not possible with all optically compatible substrates, as is silicon nitride (SI_3N_4). For this reason, and even the advantages of silicon nitride for optical biosensing applications (low water absorption at visible and near-infrared wavelengths, compatibility with CMOS processes, etc. [13]), we considered silicon as the best candidate. Using SOI alone it is thus possible to create a technology that enables electrochemical control over the surface of the sensor in situ with optical sensing. This approach eliminates the loss in sensitivity inherent with the inclusion of additional ITO layer, as the sensing layer of the device is interfaced directly with the analyte. Furthermore, the fabrication process remains CMOS compatible; a critical advantage for future technology translation and commercialisation.

There is, however, a conflict; the electrical and optical properties of the silicon substrate are related. For example, a highly doped substrate optimised for high electrical conductivity has a negative impact on the optical properties. This issue is attributed to the fact that the doping process increases the number of impurities in the crystalline silicon layer, which in turn increases the free carrier loss [14, 15]. This increase directly impacts on the performance of a resonant cavity and ultimately, if the doping levels are high enough, could completely inhibit propagation of the optical signal through the substrate. A solution to this tradeoff is to control the spatial distribution

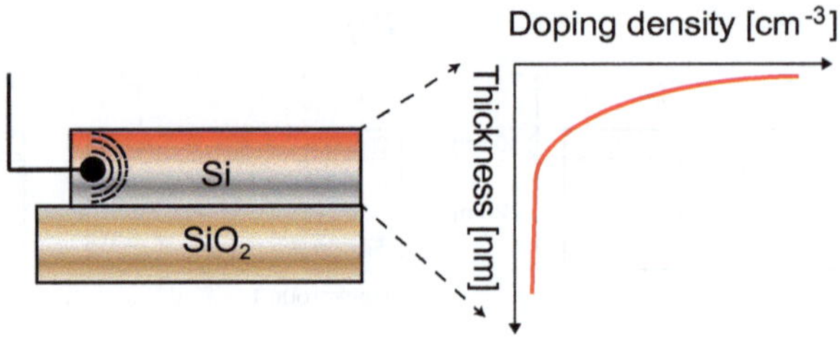

Fig. 4.2 Introduction of conductive impurities into the silicon layer of a SOI substrate. The *red line* in the silicon layer represents the doping profile and the *broken black line* represents the propagation of the electric field

of dopants in the silicon device layer (Fig. 4.2). Instead of an even doping profile through the silicon layer, I have optimised the doping process to produce a highly doped thin film located at the top surface of the silicon layer while the remainder of the silicon layer remains undoped. This is a very appealing and elegant solution. From an optical point, just the tail of the confined optical modes interacts with the thin doped layer, so the overall optical loss is minimised. From an electrochemical viewpoint, the surface layer can be sufficiently highly doped to support electron transfer to/from a solution-phase analyte.

The theory needed to understand the basics of electrochemical and photonic sensing using semiconductors has already been presented in 3.3.1. The following sections describe tuning the electrical properties of the silicon layer to meet with both the electrical and optical requirements.

4.2 Profile Controlled N-Type Doping

Given the conflict between the electrical and the optical properties of the substrate, it is clear that the doping concentration is a key parameter for the correct behaviour of the device. This section is focussed on optimising the doping concentration and doping profile in the silicon layer, which is a key parameter in the design of our biosensor. A number of characterisation techniques were employed to characterise in detail this doping from electrical, electrochemical and photonic points of view. It was decided to dope the silicon device layer *n*-type as a given electrical conductivity can be obtained with lower impurity concentration compared to *p*-type doping. As discussed in Sect. 3.2.1, the doped samples were fabricated by diffusion doping, which allows the incorporation of dopant molecules into the silicon layer at high temperature by means of a phosphorous gas flow. While ion implantation is the most commonly used approach to Si doping, implantation leads to significant damage of the lattice structure, some of which may persist even after annealing, which results in an increased optical loss of our system. However, in diffusion doping, optimising the phosphorous gas flow, temperature and exposure time, the incorporation of dopants can be controlled to produce a non-uniform doping distribution [16], as a natural result of Fick's law.

Typically, diffusion doping is used to produce an almost even doping profile in the silicon layer [17]. However, by optimising the phosphorous gas flow, temperature and exposure time, the incorporation of dopants can be controlled to produce a non-uniform doping distribution [16].

Here, the diffusion doping process was optimised to create a concentration profile with a sharp gradient i.e. a highly doped thin film in the top of the silicon layer, with ideally no dopants being present in the bulk device layer. This profile ensures a depletion region thin enough to ensure electrons can tunnel between the silicon electrode and the analyte, while minimising optical loss. The formation of a highly doped surface layer also simplifies the fabrication of the Ohmic contacts to the silicon electrode. The optimised doping process is outlined in Fig. 4.3. This process includes the required cleaning and native oxide removal steps required to obtain optimal electrical performance.

In consequence, we optimised the doping step of our samples to obtain the desired impurity gradient adjusting the gas flow, temperature program and exposure time. We found optimal results (sharp doping gradient) allowing both electrochemical measurements and fabrication of ohmic contacts, without forgetting the constraint of having a low optical loss (as shown in Sect. 4.5), with the process described in Fig. 4.3. This process includes the required cleaning and dioxide removal steps to obtain optimal electrical performance.

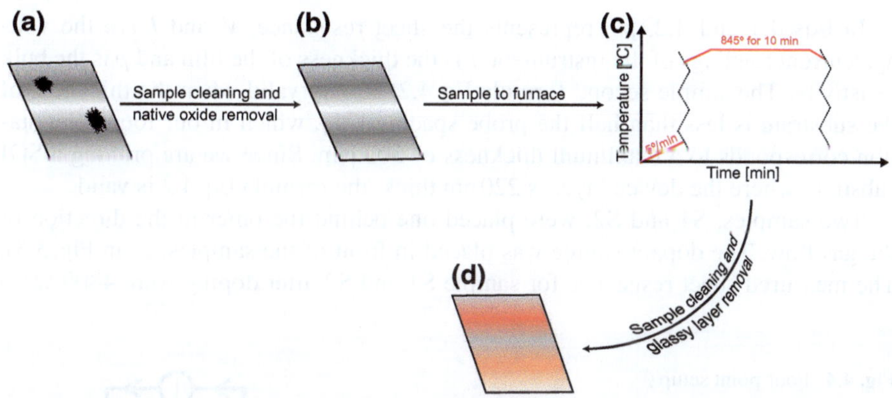

Fig. 4.3 Profile controlled n-type doping process. **a** The sample is piranha cleaned for 10 min before the native oxide layer is removed by dipping the sample in diluted HF (1:6 in DI water) for 30 s. **b** Sample and dopant source are placed on a fused quartz boat and introduced into the furnace tube under a N flow of 0.2 sccm. **c** Temperature program to achieve the desired doping profile. **d** Once cooled, the sample is again piranha cleaned and dipped into dilute HF (1:6 in DI water) for 10 s to remove the thin glassy product obtained in the *top* of the silicon layer from the doping process (deglazing)

4.2.1 Bulk Doping Characterisation

The overall doping concentration of our silicon device layer was characterised by measuring the sheet resistance, and normalising to the measured thickness of the device layer to obtain the resistivity of the film. Finally, the doping density is estimated by evaluating the mobility model of Klaaseen [18, 19] for a n-type silicon substrate. The sheet resistance was measured using a conventional a four-point probe method. This measurement of resistance is commonly employed in the semiconductor industry and relies on separate pairs of current-carrying and voltage-sensing electrodes as illustrated in Fig. 4.4. Current is supplied by an external source and a voltage drop is generated and measured across the contact points.

Considering the current and voltage reading from the probes [20], the sheet resistivity can be estimated as Eq. 4.1, while the bulk resistivity, which considers the thickness of the film, is given by Eq. 4.2. Consequently, it is possible to utilise this technique, together with Eq. 4.2, to estimate the overall doping density of our in-house fabricated samples. We measured two samples, S1 and S2, obtained from the same fabrication batch (process shown in Fig. 4.3).

$$\rho_\square(\frac{\Omega}{\square}) = \frac{\pi}{ln(2)}\frac{V}{I} \tag{4.1}$$

$$\rho = \frac{\pi}{ln(2)}t\frac{V}{I} \tag{4.2}$$

In Eqs. 4.1 and 4.2, $\rho_\square$ represents the sheet resistance, V and I are the voltage/current readings of the instrument, t is the thickness of the film and ρ is the bulk resistivity. The simple second formula Eq. 4.2, is only valid when the thickness of the substrate is less than half the probe spacing [21], which in our four-point station corresponds to a maximum thickness of 500 μm. Since we are probing a SOI substrate where the device layer is 220 nm thick, the formula Eq. 4.2 is valid.

Two samples, S1 and S2, were placed one behind the other in the direction of the gas flow. The dopant source was placed in front of the samples, as in Fig. 3.3). The measured sheet resistance for sample S1 and S2 after doping were 4800 Ω/□

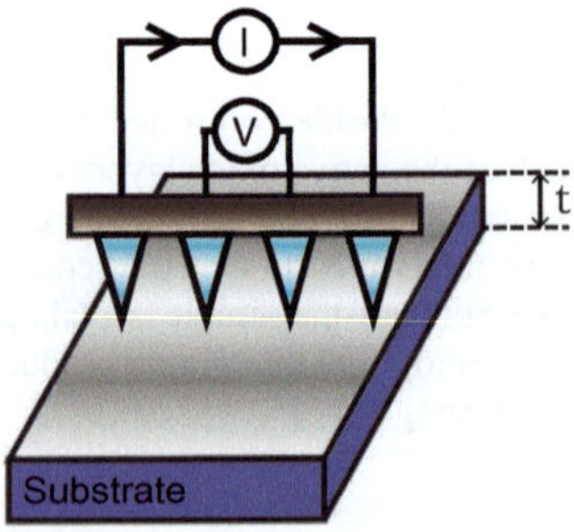

Fig. 4.4 Four point setup station schematic. Current flows between the outer probes while the inner ones sense the induced voltage in a substrate/film with a thickness *t*

and 6200 Ω/□, respectively. These values compare well with the performance data provided by the manufacturer of the n-type dopant source (see Sect. 3.2.1). We note that the higher sheet resistance observed for S2 is in part related to the location of the samples in the furnace; diffusion of dopant molecules to S2 was inhibited by S1. The resistivity of S1 and S2, calculated using Eq. 4.2, was found to be 0.105 Ω cm and 0.135 Ω cm, respectively. Finally, the overall doping densities of the samples were estimated by [18, 19], yielding values of $7.5 \times 10^{16}\,\text{cm}^{-3}$ and $5.2 \times 10^{16}\,\text{cm}^{-3}$ for S1 and S2, respectively.

4.2.2 Doping Profile Characterisation

Using this four point probe station, we were also able to characterise the doping profile of our samples with the assistance of an in-house ellipsometer. This characterisation was achieved by measuring the sheet resistance at different depths within the silicon layer of our substrates, and extracting the thickness of each layer through ellipsometry. Ellipsometry is an optical technique for investigating the dielectric properties of thin films, which measures the change of polarization upon reflection or transmission and compares it to a model that takes into account properties of the film as the thickness or refractive index. Nevertheless, before calculating both resistivity and doping density as previously performed in Sect. 4.2.1, a consideration needs to be taken into account. As the four point setup relies on the characteristics of the electric field applied between the outer probes, it has to be considered that this electric field will have a certain penetration depth in the substrate (illustrated in Fig. 4.5) which could be as large as the thin silicon layer of our device.

This means that the instrument will provide a reading which will be an average of the resistivity measured by the whole electric field. This is very detrimental in our

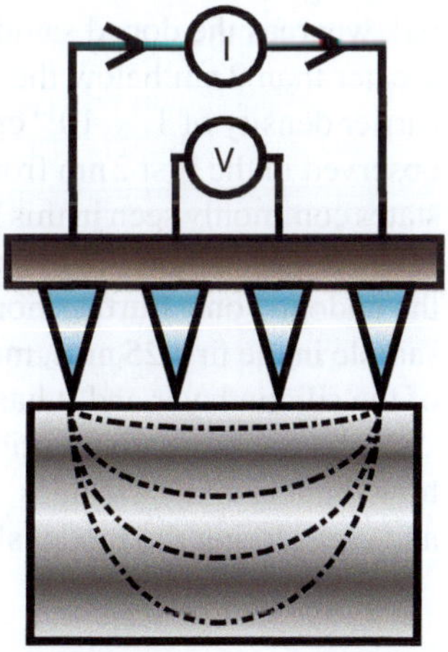

Fig. 4.5 Illustration of the electric field's penetration in a four point probe measurement station

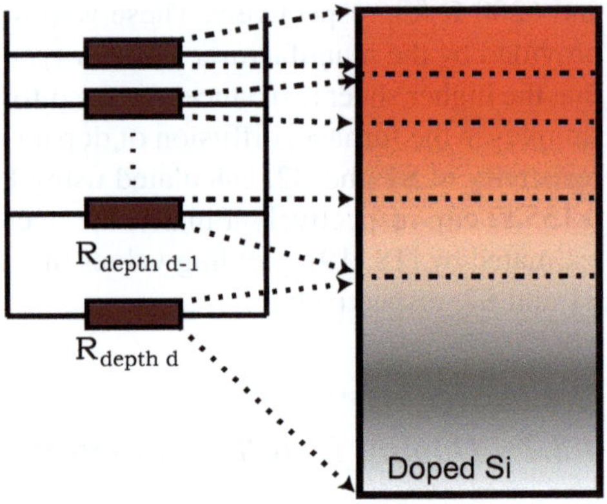

Fig. 4.6 Modelling of the resistivity of a substrate at different depths with resistances in parallel. Each of the resistances represent the resistivity of a certain layer within the film. $R_{depth\ d}$ represents the deepest resistance

scenario, as the resistivity has a gradient through the silicon substrate and we are not able to know the sheet resistance at each depth. Our solution to overcome this issue is to model the substrate as a group of resistances in parallel where each one represents the resistivity ([Ωcm]) of a layer (Fig. 4.6).

To find the resistivity of each of these layers, we sequentially dry etched the silicon layer using RIE (3.2.4) and measured the sheet resistance after each etch (as shown in Fig. 4.7). Subsequently, the resistivity of each layer was modelled as two resistances in parallel, one representing the current layer, d, and one the resistivity of the lower layer, d-1. The total resistivity is then given by: ($\frac{1}{R_{eq}} = \frac{1}{R_{depth\ d}} + \frac{1}{R_{depth\ d-1}}$). This means that we worked backwards from the resistivity of the deepest etched layer to the silicon surface. A total of three samples were measured, as shown in Fig. 4.7, two of them diffusion doped to an overall doping concentration of $7.5 \times 10^{16}\,\mathrm{cm}^{-3}$ and $5.2 \times 10^{16}\,\mathrm{cm}^{-3}$, and an undoped sample as a reference to evaluate the intrinsic doping of the substrate.

In Fig. 4.7, the intrinsic doping profile of the undoped silicon sample is shown in red, whereas the doped samples, S1 and S2, are shown in blue and green. At depths greater than 2 nm below the surface, the undoped sample is seen to have a uniform carrier density of $1 \times 10^{14}\,\mathrm{cm}^{-3}$, typical of intrinsic silicon [22]. The low resistivity observed in the first 2 nm from the surface is likely related to the presence of surface states commonly seen in this kind of substrate [23]. The sheet resistance at the surface of the doped samples is two orders of magnitude lower in the doped samples than in the undoped one. Furthermore, the sheet resistance is lower than that of the undoped sample in the first 25 nms, meaning that the doping is only present in the first 25 nms of the silicon layer and it has not diffused further in the layer as originally desired.

From the depth distribution of the sheet resistance, it is possible to also calculate how the resistivity changes with depth into the silicon device layer through Eq. 4.2 and considering the model shown in Fig. 4.6 as explained above (Fig. 4.8).

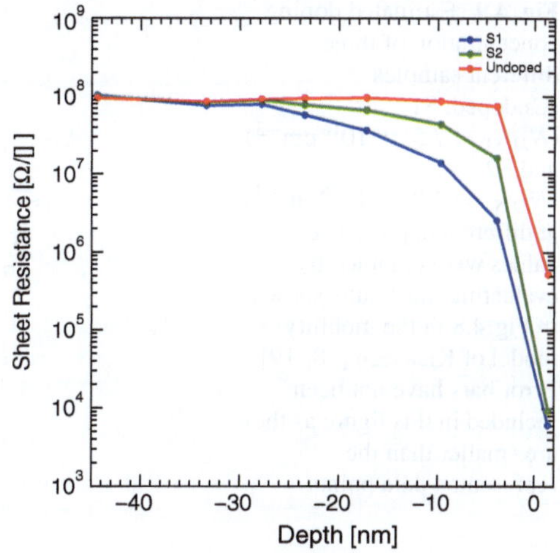

Fig. 4.7 Sheet resistance of three different samples (Undoped, S1 ($N_{D\ S1} = 7.5 \times 10^{16}\,\mathrm{cm}^{-3}$) and S2 ($N_{D\ S2} = 5.2 \times 10^{16}\,\mathrm{cm}^{-3}$) at different depths. The etch thickness was measured using ellipsometry. Error bars have not been included in this figure as they are smaller than the represented data points

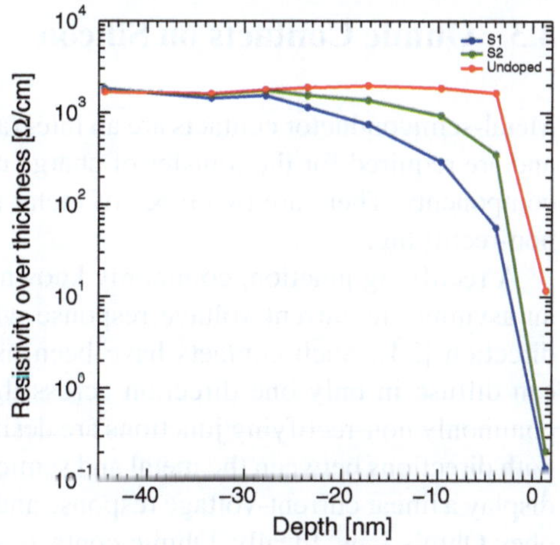

Fig. 4.8 Resistivity over thickness of three different samples (Undoped, S1 ($N_{D\ S1} = 7.5 \times 10^{16}\,\mathrm{cm}^{-3}$) and S2 ($N_{D\ S2} = 5.2 \times 10^{16}\,\mathrm{cm}^{-3}$) at different depths. The resistivity was obtained normalising the sheet resistance over the thickness of the layer and considering a group of resistances in parallel for each depth. Error bars have not been included in this figure as they are smaller than the represented data points

Subsequently, we can numerically obtain the doping density at each layer by evaluating the resistivity value in the mobility model of Klaaseen [18, 19]. The result is shown in Fig. 4.9.

As depicted in Fig. 4.9, a highly doped thin film is obtained at the top 20 nm of the silicon layer following the diffusion doping technique discussed in Sect. 4.2, which satisfies the desired doping discussed in Sect. 4.1. By integrating these doping values over the entire silicon layer, the overall doping densities are calculated to be $7.3 \times 10^{16}\,\mathrm{cm}^{-3}$ and $5.05 \times 10^{16}\,\mathrm{cm}^{-3}$, which compare well to the measured values of bulk doping density ($7.5 \times 10^{16}\,\mathrm{cm}^{-3}$ and $5.2 \times 10^{16}\,\mathrm{cm}^{-3}$, for samples S1 and S2, respectively). In Sects. 4.3 and 4.5, we validate how the device fulfils the requirements for both electrochemical and optical domains.

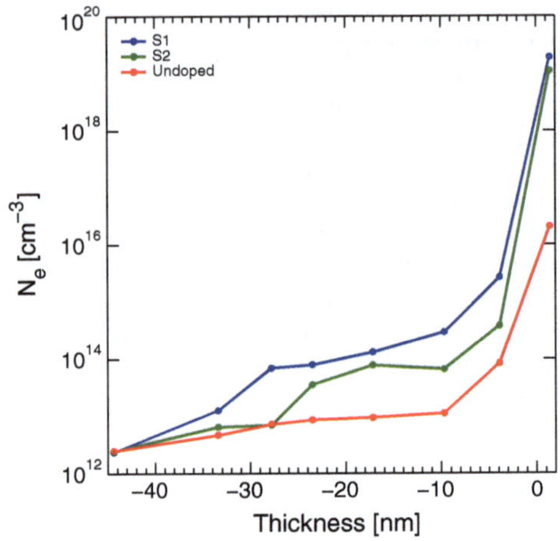

Fig. 4.9 Estimated doping concentration of three different samples (Undoped, S1 ($N_{D\ S1} = 7.5 \times 10^{16}\,\text{cm}^{-3}$) and S2 ($N_{D\ S2} = 5.2 \times 10^{16}\,\text{cm}^{-3}$) at different depths. The values were obtained by evaluating the results shown in Fig. 4.8 in the mobility model of Klaaseen [18, 19]. Error bars have not been included in this figure as they are smaller than the represented data points

4.3 Ohmic Contacts on Silicon

Metal-semiconductor contacts are an integral part of any electronic integrated device and are required for the transfer of charge carriers between the device and external components. There are two types of metal-semiconductor junctions: rectifying and non-rectifying.

A rectifying junction, commonly known as Schottky barrier, is characterised by an asymmetric current-voltage response which allows carriers to flow in just one direction [24]. Such contacts have been used to create diodes as, charge carriers can diffuse in only one direction across the metal-semiconductor interface. Most commonly non-rectifying junctions are desired where charge can move freely and in both directions between the metal and semiconductor. These non-rectifying contacts display a linear current-voltage response and are thus named Ohmic contacts, as they obey Ohm's Law. Ideally, Ohmic contacts should exhibit an extremely low contact resistance compared to the total resistance of the semiconductor material [25] and should not distort or limit the performance of the device. There exist two approaches for fabricating Ohmic contacts. One is based on lowering the height of the interfacial potential barrier through alloy compounds made of properly annealed materials [25], the other one, as exploited here, uses high doping of the semiconductor in order to minimise the barrier width to enable efficient tunnelling of electrons across the metal-semiconductor interface. This process is illustrated schematically in Fig. 4.10b.

In our case, both the Fermi level and the work function of the doped silicon layer can be calculated numerically through Eqs. 4.3 and 4.4, respectively [26].

$$E_F = E_i + kTln\frac{N_D}{n_i} \tag{4.3}$$

$$\Phi_S = (E_g + \chi) - E_F \tag{4.4}$$

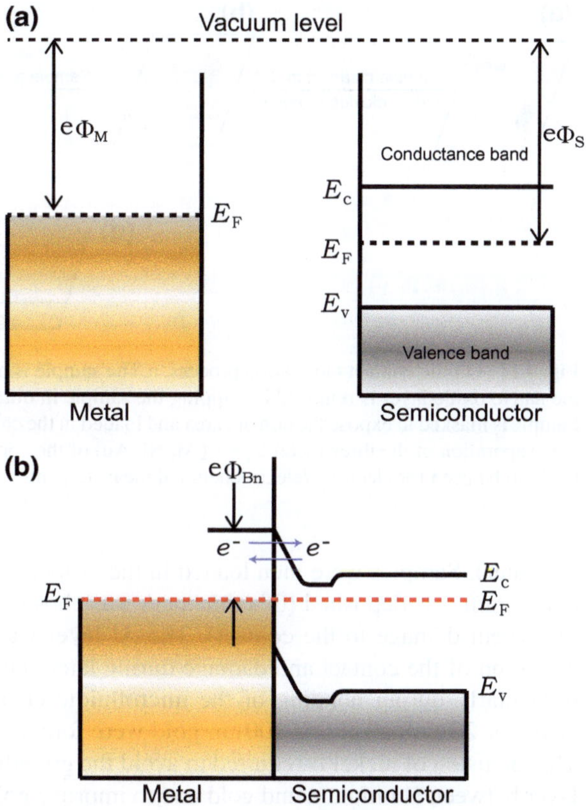

Fig. 4.10 Energy band diagram of a metal and a semiconductor before (**a**) and after (**b**) contact. Here, the interface is depicted as perfect. E_F represents the fermi level, while E_C and E_V represent the energy levels of both conductance and valence bands in the semiconductor, respectively. The work functions of metal and semiconductor are represented by Φ_M and Φ_S, respectively; while the barrier height after they are brought in contact is referred as Φ_{Bn}. In **b** the depletion region is illustrated at the metal-semiconductor interface

In Eq. 4.3, E_i, k, T, N_D and n_i are the intrinsic Fermi level, Boltzmann constant, temperature, doping concentration and intrinsic carrier density respectively. The band gap energy of the semiconductor is represented by E_g. χ is the electron affinity (energy required to remove an electron from the bottom of the conduction band to the vacuum level) in Eq. 4.4. Assuming room temperature (T = 300 °K), E_i for silicon is equal to 0.56 eV measured relative to the top of the valance band, kT=0.0258 eV, $n_i \approx 1\text{x}10^{10}\,\text{cm}^{-3}$, E_g=1.12 eV and χ=4.05 eV.

Considering $N_D = 7.5 \times 10^{16}\,\text{cm}^{-3}$, E_F equals to 1.28 eV. Thus $\Phi_{highly\ doped}$ is 3.89 E_V. Therefore, in order to obtain an ohmic contact to the silicon, we need to chose a metal material which shows a similar (or smaller) work function energy level. This is the case of aluminium (Al), whose work function has a value of 4.08 eV and which has been widely employed for this propose [27].

Thermal evaporation (described in Sect. 3.2.2) was used to fabricate the Al Ohmic contact to our device. Before loading the samples into the evaporator, the samples were cleaned and the native oxide removed to eliminate any insulating barrier at the metal-semiconductor interface. Cleaning consisted of a 10 min piranha clean (H_2SO_4:H_2O_2 mixture at 7:3 ratio) followed by a 30 s dip in diluted HF (1:6 in

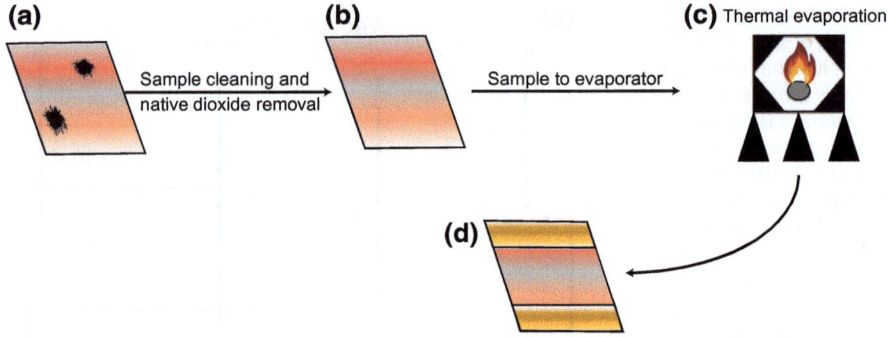

Fig. 4.11 Ohmic contact fabrication process. **a** The sample is piranha cleaned for 10 min and then the native oxide layer is removed by dipping the sample in dilute HF (1:6 in DI water) for 30 s. **b** Sample is masked to expose the contact area and loaded in the chamber for evaporation. **c** Following the evaporation of the three metal layers (Al, Ni, Au) of the contact, the sample is unloaded and is ready to be used for electrical/electrochemical measurements

DI water). Samples were then loaded in the evaporator and a 200 nm thick layer of aluminium was deposited (this thickness was chosen to provide enough robustness to prevent damage to the contact). The Al layer was further protected to prevent oxidation of the contact and damage during later stages of the fabrication process, particularly during bonding of the microfluidic channel (Sect. 3.2.5). Evaporated layers of 20 nm nickel and 200 nm gold were found to be suitable for this purpose. The inclusion of nickel is required to avoid the growth of an insulating intermetallic layer between aluminium and gold and to improve gold adhesion [28]. This contact fabrication process is illustrated in Fig. 4.11.

Fig. 4.12 shows the current-voltage characteristics of a pair of fabricated contacts. Contacts were formed on a sample with a bulk doping density of $7.5 \times 10^{16}\,cm^{-3}$ and the same doping profile as presented in Sect. 4.2. The I-V characteristics of the fabricated contact are linear across the entire voltage range, characteristic of Ohmic behaviour (i.e. no rectification at the interface). This behaviour means that electrons coming from the metallic probes can efficiently tunnel into the silicon layer through the deposited contact pads. The resistance of the contact is 327 Ω, which is calculated from the gradient of the current-voltage relation shown in Fig. 4.12. The resistance of the doped substrate (96 Ω) is subtracted to the current-voltage relation (750 Ω), and then the result is divided by two to account for both contacts.

Having demonstrated Ohmic contacts, it was also possible to further characterise the doped silicon device layer by performing Hall effect measurements enabling a direct measure of both the carrier density and mobility [29, 30]. Hall measurements were performed using the van der Pauw method which requires a sample with four Ohmic point contacts placed at the periphery of the sample (as shown in Fig. 4.13). Figure 4.14 shows the linear current voltage characteristics of the van der Pauw sample (here using a silicon substrate with a doping density of $7.5 \times 10^{16}\,cm^{-3}$ as

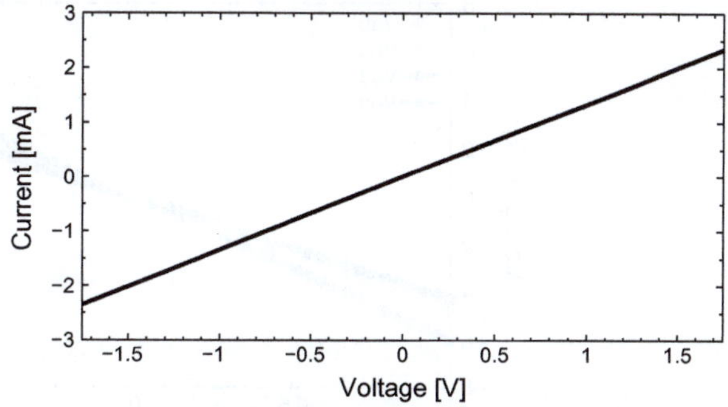

Fig. 4.12 I/V curve of a fabricated sample with a doping density of $7.5 \times 10^{16}\,\mathrm{cm}^{-3}$. Here, metallic probes are brought in contact to each of the pads and the voltage is swept from −1.8 to 1.8 V while the current is recorded. Error bars have not been included in this figure as they are smaller than the represented data points

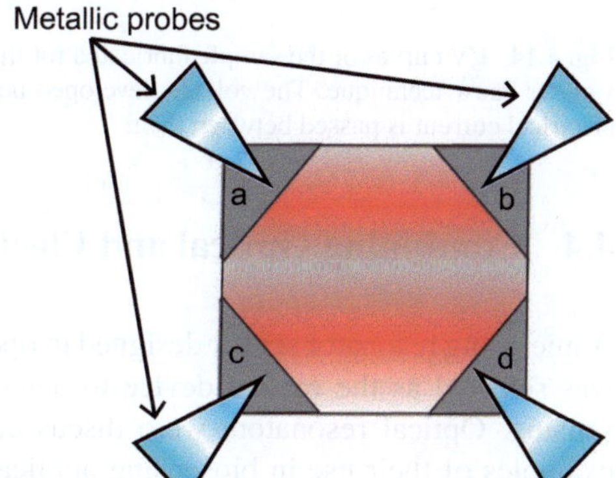

Fig. 4.13 Schematic of the van der Pauw electrode geometry employed for Hall effect measurements. Ohmic contacts with the same size are fabricated at each corner of the sample (named a, b, c and d), enabling connection to an external Hall effect measurement system

measured using the four point probe method) when current is sourced between each pair of contacts.

Briefly, the Hall effect is measured by sourcing a current between a pair of electrodes, while a magnetic field is applied in a direction transverse to the current. The Lorentz force acting on the moving charges due to the magnetic field results in the accumulation of charge on one side of the substrate. By measuring the magnitude of this potential difference (Fig. 4.14) as a function of current and magnetic field strength it is possible to determine the polarity of the doping (n-type or p-type), the doping density and the charge mobility [29, 30]. The sheet carrier density can be then calculated, and subsequently the doping concentration considering the thickness of the sample.

Here a Hall effect measurement was performed on a sample diffusion doped to a density of $7.5 \times 10^{16}\,\mathrm{cm}^{-3}$ as measured using the four point probe method. Hall measurements yielded a bulk carrier concentration of $-6.5 \times 10^{16}\,\mathrm{cm}^{-3}$ (where the negative polarity relates to the n-type doping). The electron mobility was calculated to be $88.6\,\mathrm{cm}^2\mathrm{V}\mathrm{s}^{-1}$ and the sheet resistance was $7620\,\Omega/\square$. These values are in good agreement with those presented in Sect. 4.2.2.

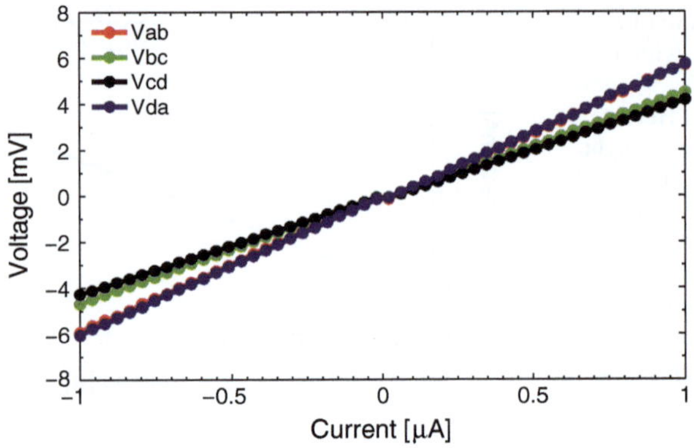

Fig. 4.14 I/V curves of the sample fabricated for the Hall effect measurement system based on the van der Pauw technique. The voltage developed between each pair of contacts is tracked when a electrical current is passed between them

4.4 Combining Optical and Electrochemical Sensing

A microring resonator sensor designed to operate in the wavelength band of 1550 nm was selected as the optical device to demonstrate proof-of-principle multi-modal sensing. Optical resonators were discussed in Sect. 2.3.2.3, together with recent examples of their use in biosensing applications. The selection of these microring sensors as an exemplar was motivated by two key advantages. Firstly, the modelling of the sensor/solution interface is relatively simple, secondly they exhibit a high tolerance to fabrication errors compared to other optical structures (e.g. photonic crystals). An example of the fabricated device using our in-house doped SOI substrate (including an Ohmic contact) is shown in Fig. 4.15.

The ring resonator and access waveguide were fabricated into the diffusion doped SOI substrate using electron beam lithography (Sect. 3.2.3) and reactive ion etching (Sect. 3.2.4). Resonator fabrication was performed following Ohmic contact fabrication (Sect. 4.3) on a profile controlled doped substrate (Sect. 4.2). The lithography and etch steps are illustrated in Fig. 4.16 and a detailed explanation and further information about these techniques/steps can be found in Sects. 3.2.3 and 3.2.4.

In the process shown in Fig. 4.16, the lithography and etching steps were optimised for our application. In order to make electrical contact to the resonator structure, It is necessary to include a continuous, conductive layer between the Ohmic contact and the cavity. This was achieved by fabricating the devices with a rib waveguide profile instead of a deep etched profile as illustrated in Fig. 4.17.

The dimensions of the rib waveguide profile were optimised via simulation to minimise leaking of the optical mode into the slab. Similarly, the physical dimensions of the ring resonator were designed to maximise the Q-factor; the Q-factor is critical to accurately monitoring low-concentration binding events. For this purpose, an all-pass ring configuration was selected (Sect. 2.3.2.3) as these exhibit a higher Q-factor

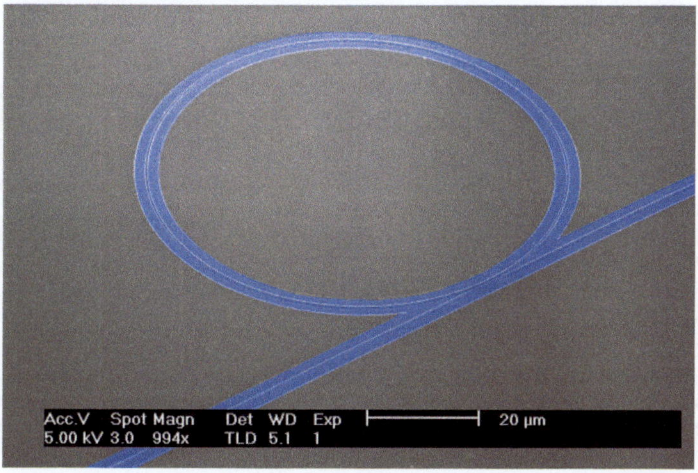

Fig. 4.15 False coloured SEM picture of a ring resonator sensor fabricated on a diffusion doped SOI substrate. The ring cavity and access waveguide are presented in *blue*, while the non-lithographed substrate remains in *grey*

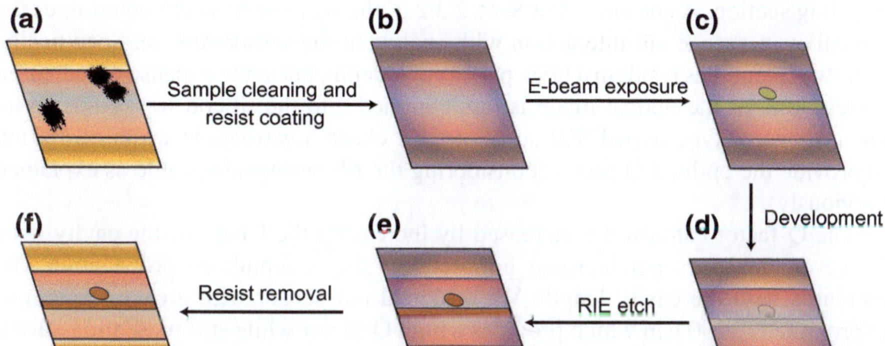

Fig. 4.16 E-beam lithography and etch fabrication steps. **a** The sample is piranha cleaned for 10 min and a layer of e-beam resist is spun at 1750 rpm and baked for 10 min at 170 °C, resulting in a 350 nm thick resist layer. **b** The optical structures are exposed on the resist by means of e-beam lithography (step size of 8 nm and a base dose of 155 $\mu C/cm^2$ at 50 kV). **c** The exposed areas are developed for 100 s in Xylene thus these patterns can be transferred into the silicon layer of the device. The sample was dry etched through RIE for 85 s in an atmosphere of 1.16:2 CHF_3:SF_6 (**d**). Finally, the resist is striped off (**e**) and the sample is washed to remove any resist residue

than devices in the add-drop configuration. The origin of this high Q-factor is the smaller roundtrip loss in all-pass rings compared to add-drop rings, and allows a higher confinement of the optical signal in the cavity and ultimately sharpens the resonance [31].

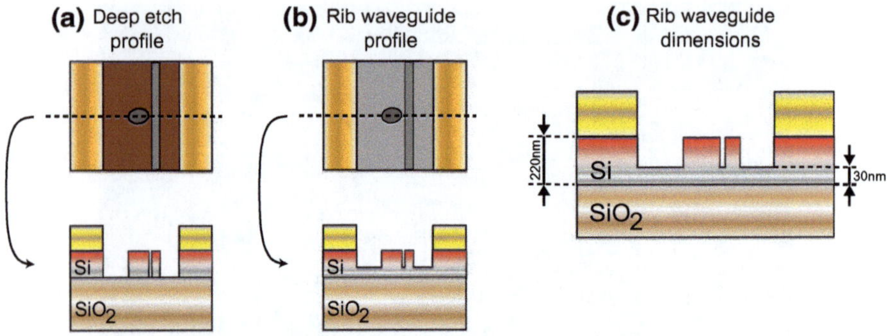

Fig. 4.17 Comparison between deep etch (**a**) and rib waveguide (**b**) profiles ring resonators. In **c** the dimensions of the rib waveguide dimensions of our ring resonator are shown. There micro ring sensors are fabricated with a rib waveguide profile to allow electron transfer from the contact to the sensing cavity

To obtain such a high Q-factor, the physical dimensions were optimised considering three design parameters: waveguide width, cavity length and gap distance of the coupling section. As discussed in Sect. 2.3.2.3, the waveguide width determines the optical evanescence tail interaction with analyte at the sensor surface. Specifically, narrow waveguides result in a large penetration depth, but at the expense of increased optical loss, as the optical mode is less confined into the silicon waveguide. With our in-house *n*-type doped SOI substrate, we chose a waveguide width of 480 nm to provide the optimal Q-factor (considering the rib waveguide profile as explained previously).

The Q-factor can also be increased by increasing the length of the cavity [31]. However, this does not increase infinitely, as the accumulated propagation loss increases with the cavity length. We designed our sensor with a cavity length of approximately 200 μm which provides a high Q-factor while still presenting a footprint on the micrometre-scale.

Finally, the distance between the optical cavity and the access waveguide (coupling section) plays a crucial role in determining the Q-factor of the cavity. Specifically, it defines the amount of optical signal that couples into (and out of) the cavity, and therefore the loss between the coupled and the uncoupled signal. Small coupling sections allow weakly coupled resonances to travel in and out of the cavity, therefore widening the range of frequencies that interfere destructively (lower Q-factor). On the other hand, large coupling sections will decrease the extinction ratio of the transmission spectrum, which is detrimental to the Q-factor. After measuring a devices with different coupling section dimensions from 150 to 300 nm, an optimal value for this coupling section was found to be 230 nm. Figure 4.18 shows the transmission spectrum around the resonant wavelength of our sensor measured in air after optimising the gap of the coupling section, while an accurate, detailed profile of the geometry of the ring resonator (including substrate) is given in Fig. 4.19.

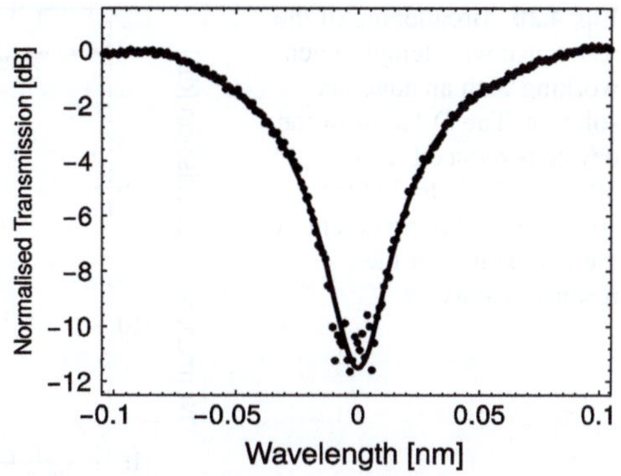

Fig. 4.18 Spectral performance of our ring resonator sensor after optimising the gap of the coupling sections. A maximum Q-factor value of 50,000 is found for a gap distance of 230 nm, using our in-house doped SOI substrate

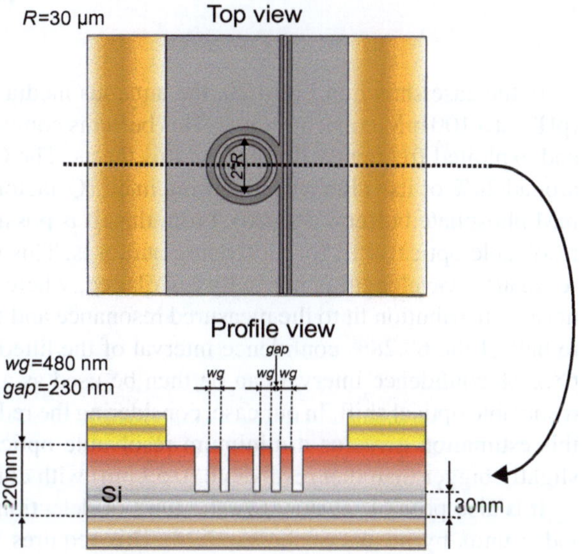

Fig. 4.19 Detailed cross-profile figure of the device. A ring cavity length of approximately 200 μm (R = 30 μm), a waveguide width of 480 nm and a gap distance of 230 nm were found optimal to obtain a maximum Q-factor of 50,000

As can be calculated from picture Fig. 4.18, the Q-factor is of the order of 50,000 for our device fabricated with doped silicon. This is comparable to ring resonator sensors fabricated in industrial foundries and published in the literature [31–34]. When working in aqueous media (i.e. when a buffer solution is flowing over the sensor surface), the resonant peak broadens due to increased loss in the resonant cavity. This increased loss is due to a combination of water absorption at our working wavelength band (around 1550 nm), and the change in the coupling coefficient. The change in the coupling coefficient is originated from the change in the refractive index on the sensor surface, as the mode confinement decreases with the reduction in refractive index contrast upon exposure to aqueous media. The broadening of the resonance wavelength by comparing the size of the resonance dip before after is shown in Fig. 4.20.

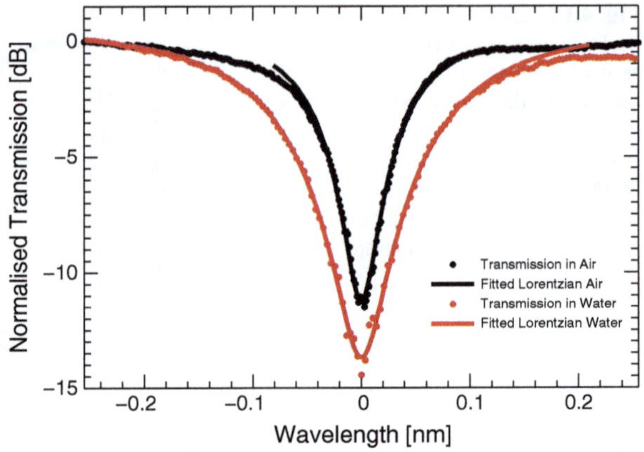

Fig. 4.20 Broadening of the resonance wavelength when working with an aqueous solution. The Q-factor of the device is reduced from 50,000 (*black*) to 38,000 (*red*) due to the increment in the optical loss of the resonant cavity

In the case shown in Fig. 4.20, the aqueous media was sodium phosphate buffer (pH7, at a 100 mM concentration). This buffer is commonly used in biological assays and exploited frequently throughout this thesis. The Q-factor is found to reduce by around 24% of its value when working in air (Q-factor air = 50,000, Q-factor in 100 mM phosphate buffer = 38,000). From this, it is possible to calculate the minimum resolvable optical shift for biosensing purposes. This is estimated assuming that the resonance wavelength is normally distributed, where the mean is identified from a normal distribution fit to the measured resonance and the standard deviation is equal to half of the 68.28% confidence interval of the fitted distribution [35]. Half of the 68.28% confidence interval can be then be used as an estimation of the minimum resolvable optical shift. In our case, considering the red resonance shown in Fig. 4.20, this estimation gives us a minimum resolvable optical shift of 0.75 pm, which is slightly bigger than that resolvable (0.59 pm) with commercial devices [36].

It is also possible to quantify the limit of detection (*LOD*) in terms of refractive index units by means of the Eq. 2.18. This requires knowledge of the bulk sensitivity of our device, given in [nm (resonance shift)/RIU (refractive index units)] (Sect. 2.3.2.3). The sensitivity was calculated empirically by tracking the resonance wavelength of the device when challenged with a range of NaCl solutions of different molar concentrations, and thus different refractive indexes. The refractive index of each solution was calculated through the expression $n_{NaCl} = 1.3105 + 0.17151 \times C$, where C is the concentration of the NaCl solution in mass % [35]. Figure 4.21 shows the result of this experiment and displays a clear linear relationship between resonance shift and solution concentration, where the gradient of the line of best fit was 65.1 nm/RIU. A bulk sensitivity of 65.1 nm/RIU, together with the minimum resolvable optical shift of 0.75 pm, leads to a *LOD* of 1.15×10^{-5} RIU. This value is in good agreement with similar ring resonators presented in the literature [31, 32, 35].

Regarding the reproducibility of the fabrication process, I found that the main source of fabrication deviations was the stability at high temperatures in the doping process. Our furnace temperature control unit presents a control down to 1°C. This

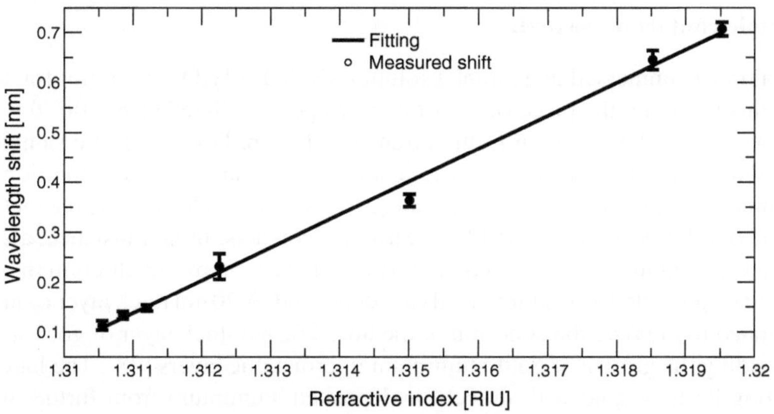

Fig. 4.21 Resonance wavelength shift against changes in the bulk refractive index. A sensitivity of 65.1 nm/RIU is obtained (equivalent to a *LOD* of 1.15×10^{-5} RIU), which shows that our sensitivity is more than sufficient to perform accurate biosensing experiments

allowed us to obtain reasonably high reproducibility between fabrication batches, for example, the maximum batch to batch deviation in the averaged sheet resistance (at a depth of 0 nm and considering three points throughout the sample) was 100 Ω/□. This, compared to the usual value of 4860 Ω/□ results in maximum fabrication deviation of 2%.

4.4.1 Summary of the Sample Fabrication Process

In this section a summary of the sample fabrication process is given. An in-depth discussion of the the nanofabrication processes here employed are presented in Chap. 3.

1. SOI doping

A piece of 2 × 1.54 cm of SOI is cleaved from a 8"master SOI (100) wafer (SOITEC, France). The wafer consists of a 220 nm-thick layer of silicon on top of a 2 mm layer of buried oxide. The sample is piranha cleaned for 10 min before the native oxide layer is removed by dipping the sample in diluted HF (1:6 in DI water) for 30 s. Sample and phosphorus dopant source (Phosphorus Grade PH-950, Saint-Gobain Ceramics, USA) are placed on a fused quartz boat and introduced into the furnace tube under a nitrogen flow of 0.2 sccm. The temperature was ramped up to 845 °C at 5 °C/min, where it was maintained for 10 min. Once cooled, the sample is again piranha cleaned and dipped into dilute HF (1:6 in DI water) for 10 s to remove the thin glassy product obtained in the top of the silicon layer from the doping process (deglazing).

2. Ohmic contact deposition

The sample is immersed in a piranha solution ($H_2SO_4 : H_2O_2$, 7:3 ratio) for 10 min. After this cleaning, the piece of substrate is dipped in diluted (1:6 ratio) for 30 sec to remove oxide layer grown in the piranha etch step. Following, the sample is is masked to expose the contact area and loaded in the chamber for evaporation in the minimum amount of time to minimise the grown of any oxide layer. Three metals are deposited to fabricate robust and low resistance contacts. In first instance, a 200 nm thick layer of aluminium is evaporated. As for further steps the aluminium contact needs to be protected, two more metals are deposited. A 20 nm thick layer of nickel is evaporated to improve the adhesion of the third 200 nm thick layer of gold, and also for avoiding the growth of gold-aluminium intermetallic layers [28]. The latter non-oxidizing layer of gold will protect the deposited aluminium from further piranha cleaning steps.

3. Electron beam lithography and etch

The sample is piranha cleaned for 10 min and a layer of positive e-beam resist(AR-P 6200.09, Allresist, Germany) is spun at 1750 rpm and baked for 10 min at 170 °C. Ring resonators are exposed on the resist by means of e-beam lithography (step size of 8 nm and a base dose of 155 $\mu C/cm^2$ at 50 kV). The exposed areas are developed for 100 s in Xylene thus these patterns can be transferred into the silicon layer of the device. The sample was dry etched through RIE for 85 s in an atmosphere of 1.16:2 (CHF_3:SF_6). The resist is removed in microposist remover 1165 at 75 °C for 45 min and the sample is washed to remove any resist residue. Finally the device is cleaved, resulting in a device with final dimension of approximately 1 × 2 cm.

4. Microfluidics mold bond

A previously fabricated piece PDMS (Sect. 3.2.5) with a 100 × 200 μm channel is cut to size. Following the punching of 2 mm holes to access the microchannel, it is sonicated while dipped in IPA for 5 min. To bond the PDMS block to the SOI chip, oxygen plasma was used for creating free OH groups on the PDMS surface. After activation, the PDMS block is brought into contact and aligned with the SOI sample to from a covalent siloxane (Si-O-Si) irreversible bond. Note that in parallel, the SOI sample has been piranha cleaned for 10 min to ensure that the surface is free from organic material, whilst adding OH groups on the silicon surface. Finally, it is placed in a 135 °C for 3 h to obtain a strong bond. Figure 4.22 shows an image of the finished sample.

4.5 Doping Impact on the Optical Device Performance

Optical loss is the major concern in doped silicon photonic devices; high doping will increase free-carrier losses and limit the performance of the device in the optical regime. The impact of the optical loss is twofold. Firstly, in resonant cavities, it will

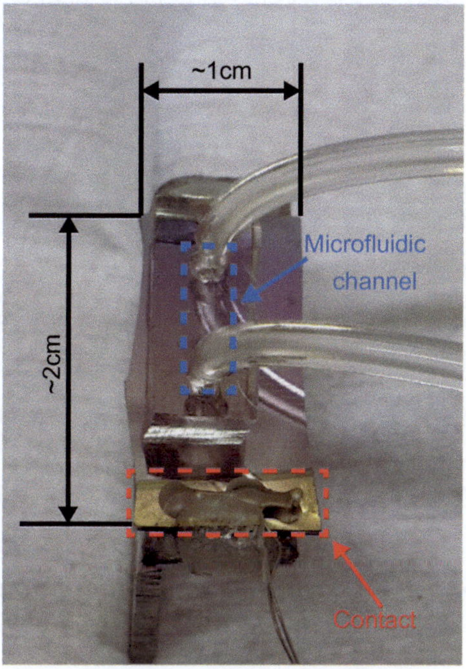

Fig. 4.22 Picture of the final device. The final device has dimensions of approximately 1×2 cm. A close-up image of the microfluidic channel is shown in Fig. 3.9

increase the loss per roundtrip, and therefore decrease the Q-factor. Secondly, very high optical losses will lead to complete absorption of the light, so no signal can be detected at the end of the waveguide. As discussed in Sect. 4.2, the SOI substrate was modified with a doping profile designed to minimise the impact on the optical performance of the device. This section explores the impact of diffusion doping on the performance of the ring resonator optical sensor.

Figure 4.23, compares the resonance wavelength before and after the diffusion doping of the substrate. Here, the substrate was doped using the processes defined in Sect. 4.2 such the average dopant concentration in the device layer of the SOI substrate was 7.5×10^{16} cm^{-3}.

The Q-factor of the doped device is marginally smaller than observed for the undoped substrate (50,000 compared to 65,000 respectively). This supports the hypothesis that the losses can be minimised using profile controlled doping. In the next section, we analytically asses this broadening to further validate the profile controlled doping fabrication process.

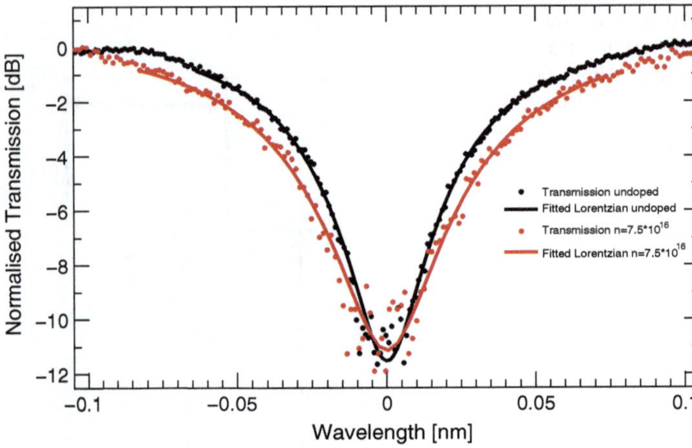

Fig. 4.23 Moderate broadening in the resonance wavelength due to the increase in optical loss. The Q-factor of the undoped resonance (*black*) is 65,000, while the doped one (*red*) is 50,000

4.5.1 Assisted Optical Validation of the Q-Factor of a Lossy Cavity

As defined in Sect. 2.3.2.3, the Q-factor of a cavity is proportional to the energy stored in the cavity divided by the energy lost by cycle. This can be formulated as Eq. 4.5:

$$Q = 2\pi \times \frac{Energy\ stored}{Energy\ lost/cycle} = 2\pi \times \frac{U_0}{U_0(1 - e^{-\alpha L})} \tag{4.5}$$

In Eq. 4.5, α is the optical absorption (directly related to free carrier loss) at the working wavelength, L the length required to complete a single optical cycle, and U_0 is the energy stored. For clarity, please note that Q refers to the loss per optical cycle, not per cavity roundtrip. Hence, L can be rewritten as a function of the wavelength as $L = \lambda/n_{eff}$, such that Eq. 4.5 is reformulated as:

$$Q = 2\pi \times \frac{1}{1 - e^{-\alpha L}} = 2\pi \times \frac{1}{1 - e^{-\alpha \frac{\lambda}{n_{eff}}}} \approx \frac{2\pi}{\alpha \frac{\lambda}{n_{eff}}}\ (Using\ e^{x} = 1 + x,\ for\ small\ x) \tag{4.6}$$

In Eq. 4.6, λ is the wavelength and n_{eff} is the effective refractive index of the cavity. The optical absorption, α, which is proportional to the doping density, can be obtained by the method described in [37]. The dependence of optical absorption on doping density is shown in Fig. 4.24.

At the working wavelength (1550 nm), the optical absorption assuming an overall doping density of $7.5 \times 10^{16}\,\text{cm}^{-3}$ is $4.2\,\text{cm}^{-1}$ (or $4.2 \times 10^{-5}\,\mu\text{m}^{-1}$). Here, the

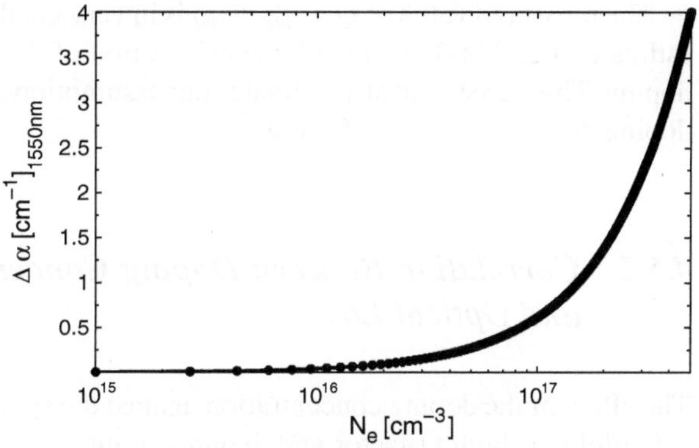

Fig. 4.24 Optical absorption for n-type doping at 1550 nm. Absorption exponentially increments with the 10-log of the doping density at our working wavelength [37]

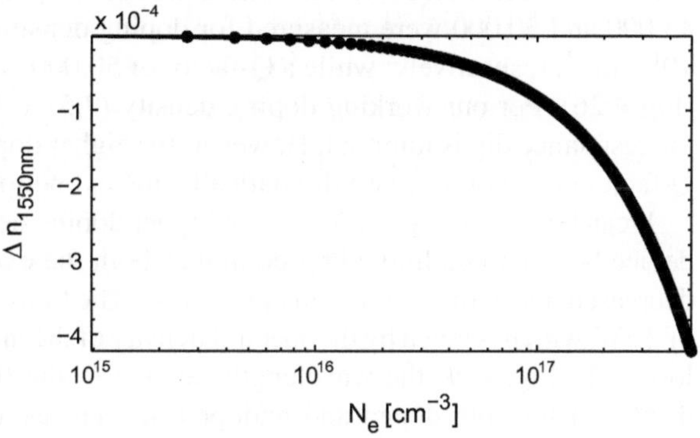

Fig. 4.25 Change in the silicon refractive index due to n-type doping at 1550 nm. A marginal change of -6×10^{-5} is appreciated at our doping density [37]

doping is assumed to be distributed evenly throughout the device layer. The refractive index of the silicon is also modified by doping, due to the presence of impurities (here phosphorous). This change in refractive index can also be modelled as described in [37] and plotted in (Fig. 4.25).

For a doping density of 7.5×10^{16} cm^{-3} as used here, only a small change in the refractive index of -6×10^{-5} is expected. Assuming n_{eff} of 2.35 (value simulated through COMSOL Multiphysics at our working wavelength) and an approximated theoretical intrinsic Q_{abs} of 222,000 is obtained.

It is also possible to obtain the intrinsic $Q_{abs\ measured}$ of the fabricated devices from measurement of the doped device. From Fig. 4.23, we obtain $Q_{doped} \approx$ 50,000 and $Q_{undoped} \approx$ 65,000. Since Q_{doped} can be obtained as:

$$\frac{1}{Q_{abs\ measured}} = \frac{1}{Q_{doped}} - \frac{1}{Q_{undoped}} = \frac{1}{50{,}000} - \frac{1}{65{,}000} \tag{4.7}$$

$$Q_{abs\ measured} = 216{,}000 \tag{4.8}$$

The measured value of $Q_{abs\ measured}$ is in very good agreement with our analytical estimate of 222,000 obtained by consideration of the increased loss induced by the doping. This agreement also validates our assumption and measurement of the overall doping density of $7.5 \times 10^{16}\,cm^{-3}$.

4.5.2 Correlation Between Doping Concentration and Optical Loss

The effect of the doping concentration against the spectral performance of the device is twofold, as both Q-factor and dynamic range of the cavity are decreased. This is attributed to the increase in the optical loss due doping of the substrate. Q-factors of 45,000 and 30,000 were measured for doping densities of $1 \times 10^{17}\,cm^{-3}$ and $3 \times 10^{17}\,cm^{-3}$, respectively; while a Q-factor of 50,000 is obtained for $7.5 \times 10^{16}\,cm^{-3}$ (Fig. 4.26). For our working doping density ($7.5 \times 10^{16}\,cm^{-3}$), the broadening of the resonance dip is minimal. However, for higher doping densities, even though the Q-factor does not decrease dramatically, the extinction ratio is noticeably reduced.

It can be seen in Fig. 4.26 how, for higher doping densities, the performance of the device begins to be limited by doping as both the extinction ratio and Q-factor are decreased due to the increase in optical loss. The transmission intensity is in the order of 2 nW, which is given by the spectral density of the input source (ASE) and coupling losses. In Fig. 4.26, the wavelength relative to the fitted resonance wavelength is displayed for both doped and undoped resonances, enabling comparison between resonance dips. I note that the absolute resonance wavelength of the doped and undoped devices differs due to small variations in the fabrication process, which have a greater impact than the marginal change in the refractive index caused by doping of the device.

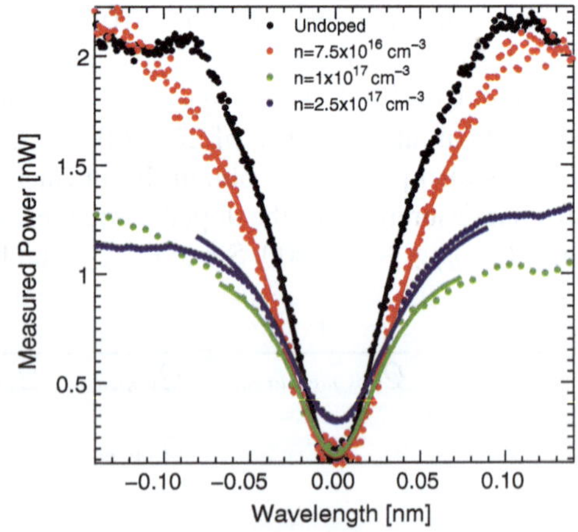

Fig. 4.26 Dependence of the Q-factor and the extinction ratio on the average doping density of the material. For our working doping density, the broadening of the resonance dip is minimal. However, for higher doping densities, even though the Q-factor does not decrease dramatically, the extinction ratio is noticeably reduced

4.6 Dual-Mode Photonic/Electrochemical Validation

The previous section quantified the impact of diffusion doping on the optical characteristics of the ring resonator sensors. This section assesses the detection sensitivity of the doped device to quantify analytes local to the sensor surface and whether electrochemical measurements (in parallel with photonic measurements) can be performed at these doping concentration levels.

In this demonstration, we monitored and quantified the immobilisation of a layer of redox active molecules on the surface of our sensor, from both electrochemical and optical domains. This molecular monolayer, shown in Fig. 4.27, consists of a long-chain alkanethiol that incorporates an oligoethyleneglycol (PEG) linker derivatised with the redox active molecule methylene blue (MB) [38]. The molecule assembles into a dense monolayer (hence the name MB-SAM) when immobilised on a surface driven by interactions between methyl groups in the alkane chain. Details about the solid phase synthesis of this molecule and its electrochemical properties can be found in [38].

The doped resonator sensor was fabricated following the processes described in Sects. 4.2, 4.3 and 4.4, before a microfluidic channel was bound to the sensor surface (Sect. 3.2.5). Finally, the complete device was placed in the electro-optic characterisation setup (Sect. 3.4). The sensor surface was required to be made thiol reactive in order to bind the MB-SAM. This surface was thus modified using silane chemistry to create a thiol surface (through an MPTS layer 2.5.2) after which an aqueous solution of Cu^{2+} ions was added. These Cu^{2+} ions bind to the surface thiol, reducing to Cu^{1+} and rendering the surface thiol reactive [39]. Finally, a 0.1 M aqueous solution of the MB molecule was passed over the sensor surface, after which the thiol group binds to the Cu^{1+} to create a dense monolayer. This immobilisation

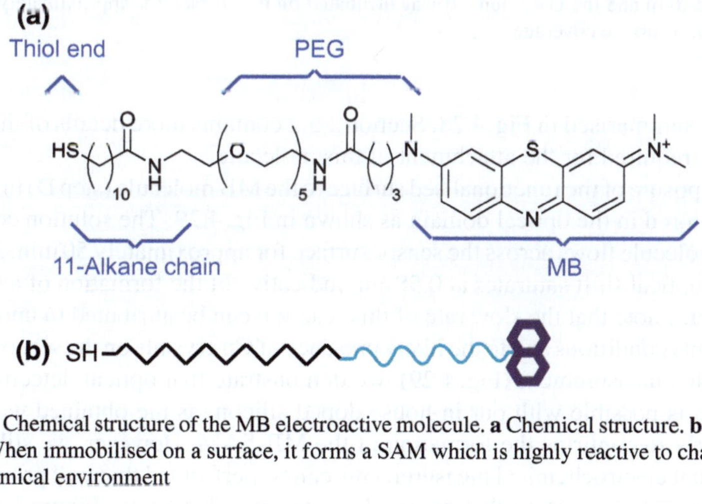

Fig. 4.27 Chemical structure of the MB electroactive molecule. **a** Chemical structure. **b** Simplified version. When immobilised on a surface, it forms a SAM which is highly reactive to changes in its electrochemical environment

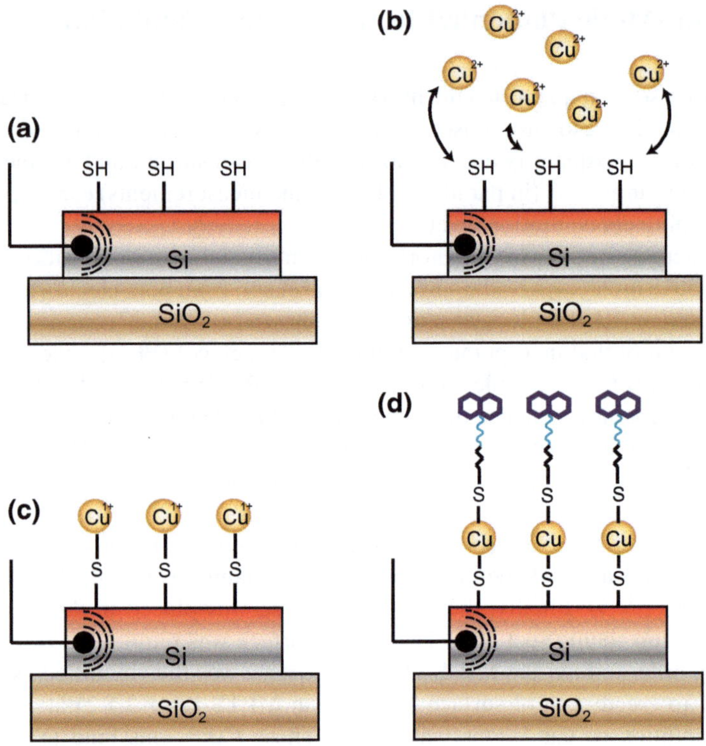

Fig. 4.28 Silane chemistry to form a MB-SAM on silicon. **a** The surface was modified with a thiol by exposure to a dilute solution of (3-Mercaptopropyl)trimethoxysilane (MPTS) through the PDMS channel. **b** Next, a 15 mM aqueous solution of Cu^{2+} ions was injected and left to incubate for 15 min. **c** These Cu^{2+}ions react with the thiol group releasing the hydrogen atom and reducing the Cu^{2+} to Cu^{1+}. **d** Finally, the MB molecule [38] is immobilised on the surface driven by a reaction between the thiol and the Cu^{1+} ions. It was incubated on the surface for approximately 50 min to achieve a high surface coverage

process is summarised in Fig. 4.28. Section 2.5.2 contains more details of the surface chemistry required for the attachment of this molecule.

The exposure of the functionalised surface to the MB molecule (step D) in Fig. 4.28 was monitored in the optical domain as shown in Fig. 4.29. The solution containing the MB molecule flows across the sensor surface for approximately 50 min. After this time, the optical shift saturates at 0.58 nm, indicative of the formation of a complete monolayer. I note that the slow rate of this reaction can be attributed to unoptimised experiments conditions and to the likely presence of Cu crystals on the sensor surface.

With this measurement (Fig. 4.29), we demonstrate that optical detection of bio molecules is possible with our in-house doped silicon, as the obtained wavelength shift clearly to confirms the formation of the MB-SAM. However, we still need to validate that electrochemical measurements can be performed that will provide complementary information to that revealed in the optical domain. Figure 4.30 shows

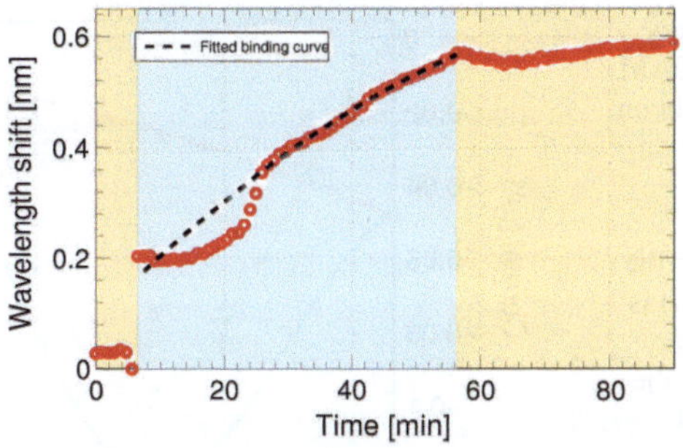

Fig. 4.29 Optical binding curve of our functionalised surface with silane chemistry upon exposure to the MB molecule. A MB-SAM monolayer is obtained after approximately 50 min from the injection of the solution (*light blue area*), providing a wavelength shift of 0.58 nm (from *light orange* to *light orange area*). The fitted curve (*black broken line*) is done to an order two polynomial function

Fig. 4.30 Electrochemical response of our functionalised sensor with a MB-SAM monolayer. Three CVs responses are shown for different doping concentrations (undoped silicon—*black*, 7.5×10^{16} cm^{-3}—*red* and 1×10^{20} cm^{-3}—*green*. All the samples were provided with ohmic contacts and the scan rate was 50 mV/s

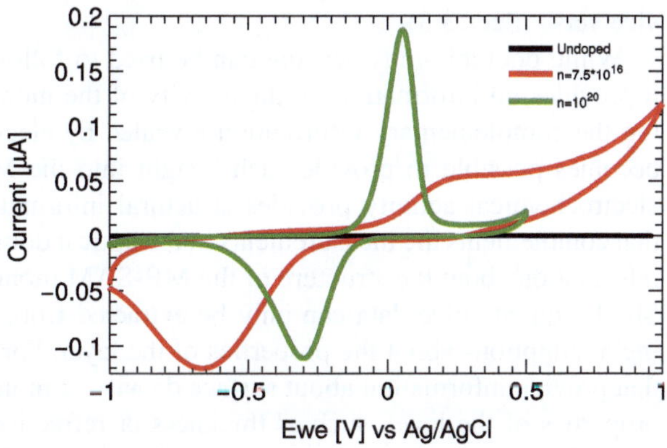

a typical cyclic voltammogram of the MB-SAM functionalised sensor. Two identically functionalised silicon samples provided with Ohmic contacts but with different doping concentrations (undoped and 1×10^{20} cm^{-3}) are also included.

Both the reduction and oxidation peak are clearly discernible for the doped ring resonator sensor, confirming the sensor is capable of electrochemical detection of redox molecules. Therefore photonic and electrochemical measurements can be done in parallel using our sensor. I note that no redox activity is observed for the undoped device. Figure 4.30 also shows the importance of doping concentration in the electrochemical measurements. Specifically, the redox peaks are more widely separated in the lightly doped silicon resonator (10^{16} cm^{-3}) compared to the heavily doped planar silicon substrate (1×10^{20} cm^{-3}). This can be understood by considering the depletion layer that forms at the silicon-electrolyte interface. The width of the depletion region is inversely proportional to the doping concentration. Thus, with low doping, a large overpotential needs to be applied to reduce the width of the depletion layer

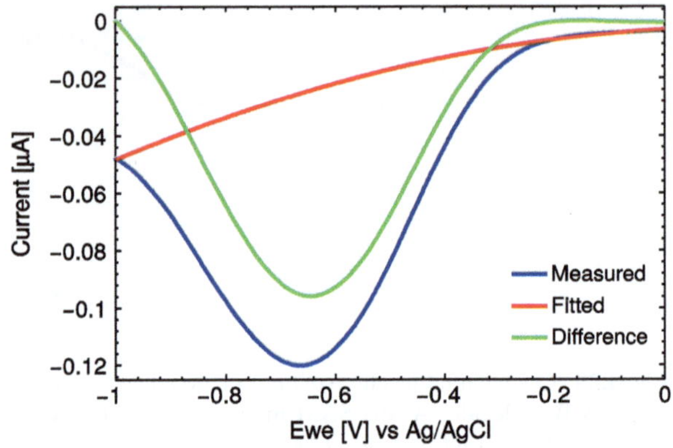

Fig. 4.31 Surface density calculation of the MB-SAM formed on our doped silicon sensor. The background current (*red*) from the reduction peak (*blue*) is removed and the area of the normalised curve (*green*) is integrated to calculate the transferred charge from the silicon into the attached molecules

before electrons can tunnel from the solution to the immobilised MB molecule and vice versa (Sect. 3.3).

While photonic-only sensing can be used to follow assembly of the MB-SAM, it provides no information on the activity of the monolayer. However, by combining the complementary information revealed by electrochemical measurements, it becomes possible to provide such insight into the MB activity. Furthermore, the electrochemical activity provides structural information regarding the monolayer that complements the measurements in the optical domain. In particular, quantitative information about the structure of the MB-SAM monolayer can be probed directly. Similar quantitative data can only be extracted from the photonic domain following assumptions about the properties of the layer. For example, literature examples that provide information about surface density of molecular layers all assume some properties of the layer, such as thickness or refractive index, that are often poorly defined [40]. In the electro-photonic sensor however, the area under the reduction and oxidation current peaks can be used to calculate precisely the total number of surface attached molecules, and thus the molecular density without making any assumptions (Fig. 4.31).

The procedure to calculate the molecular layer density of an electroactive molecule (such as the immobilised MB molecule) consists of three steps. Firstly, the background current is subtracted from the current peak to remove the capacitive charging current. The total transferred charge, C, is then given by the area under the current peak, found by integrating the curve with respect to voltage. For this integration, the scan rate (V/s) is taken into account to transform the voltage axis into the time domain. Secondly, the total charge is normalised to the electrode area exposed to the electrolyte (which in our case is given by the dimensions of the microfluidic channel) to obtain the charge density in C/cm^2. Finaly, the charge density is divided by the electron charge ($1.602 \times 10^{-19}\,C$) and number of electrons involved in oxidation and reduction (MB is a two-electron process [38]) to yield the number immobilised redox active molecules. Applying this procedure to the data presented in Fig. 4.31, the total charge transferred is $1.13 \times 10^{-8}\ C$, corresponding to a surface density of 2.2×10^{12} molecules/cm^2. This compares well with the surface density of a MB-SAM assembled on gold [41].

4.7 Limitations

Section 4.1 presented two approaches for combining electrochemical and photonic sensing using a silicon photonics platform. The optical loss that results from the addition of a conductive and transparent layer such as ITO was shown to be detrimental to the sensor performance. As a result this approach was rejected in favour of tuning the electrical properties of the silicon layer to allow dual-mode sensing. Here I discuss the limitations associated with this all-silicon alternative.

The first limitation is the electro-optic effect. The optical properties of many materials, including silicon, change in response to the application of an electric field, which typically varies slowly compared with the frequency of the light propagating through the material. When a low frequency electrical potential is applied either between the contact pads at the edge of the sample, or between one contact pad and a counter electrode in solution, electrons will flow between these points and will locally modify the optical properties of the doped material, including the optical absorption and the refractive index and permittivity [42].

In my electro-photonic device, this effect is only observed when applying a strong electric field across the device (contact pad to contact pad) for extended periods of time. Here, the resonance wavelength of the device shifts upon the application of large DC voltages above 3 V. This induced shift is shown in Fig. 4.32. In parallel, the intensity of the optical signal is also affected by the application of such large DC voltages.

Figure 4.32 depicts how after the application of 1 and 2 V for approximately 10 min, the resonance wavelength remains unchanged. However, after the applica-

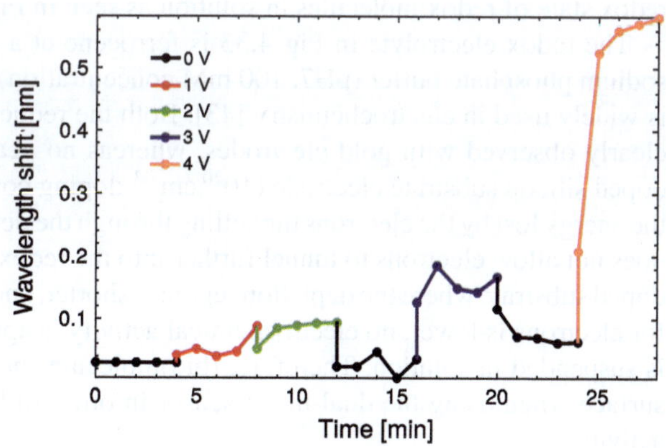

Fig. 4.32 Wavelength shift of optical resonance wavelength upon application of strong electric fields for extended periods of time. Due to the electro-optic effect, the optical properties of the material are modified, thus the resonance wavelength is affected

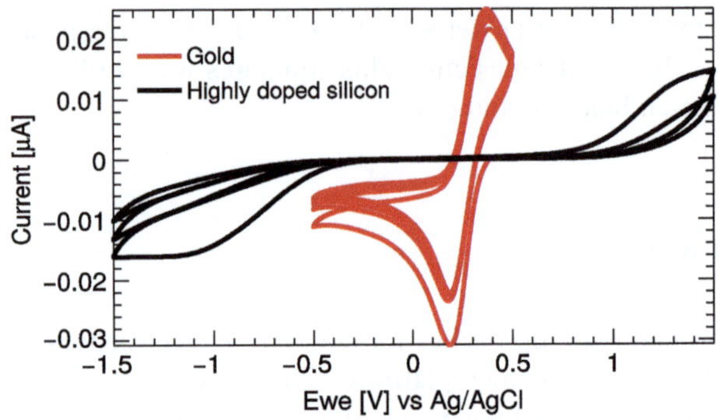

Fig. 4.33 Comparison of the electroreduction of a suspended molecule (ferrocene) in solution between gold and highly doped silicon substrates. Clear ferrocene redox peaks can be observed when using a gold substrate, but no peaks are observed with a highly doped silicon substrate ($10^{20}\,cm^{-3}$ doping concentration)

tion of 3 V for approximately 5 min, the initial resonance wavelength is no longer recovered. Finally, upon the application of 4 V, an extremely large optical shift is produced due to the electro-optic effect. It is also worth noting that upon continuous application of such high voltage values, the induced optical shift can be higher than that obtained in biosensing experiments, and therefore will interfere with the optical measurement signal. I note, that such large electrical potentials exceed the range commonly used for electrochemical measurements.

The second limitation is due to the depletion region at the silicon-analyte interface. As explained in 3.3, when a semiconductor is brought into contact with an electrolyte, a depletion region develops inside the semiconductor, which will be present even for highly doped materials. For molecules immobilised on the surface of the semiconductor, the width of this depletion region directly relates to the overpotential needed to either reduce or oxidise immobilised molecules as empirically shown in Fig. 4.30. However, while with our sensors, it is still possible to change the redox state of immobilised molecules, it has not yet been possible to modify the redox state of redox molecules in solution as seen in Fig. 4.33.

The redox electrolyte in Fig. 4.33 is ferrocene at a concentration of 5 mM in a sodium phosphate buffer (pH7, 100 mM concentration) supporting electrolyte. This is widely used in electrochemistry [43]. Both the reduction and oxidation peaks are clearly observed with gold electrodes, whereas no peaks are shown with a highly doped silicon substrate electrode ($10^{20}\,cm^{-3}$ doping concentration). This is because the energy lost by the electrons tunnelling through the depletion region at the interface does not allow electrons to tunnel further into the redox molecule. Even for a highly doped substrate where the depletion region is shorter, and therefore the drop of energy for electrons is lower, no electrochemical activity is appreciated when the molecule is suspended in solution. Therefore, (bio)molecules need to be immobilised on the surface when using the dual-mode sensor in order to look into their (bio)chemical activity.

Having shown both advantages and limitations of our approach, a comparison between similar bi-domain approaches is given in the next section. Our proposed sensor is compared to other optical and mechanical dual-mode sensing techniques.

4.8 Comparison to Similar Bi-Domain Approaches

A number of "single-domain" biosensors have been demonstrated that exploit electrical mechanical or optical methods to transduce binding events (see Sect. 2.3). In contrast, the range of solutions developed for multi-domain sensing is significantly more limited (2.4). This section will compare the most widely known bi-domain sensors, namely Electrochemical Quartz crystal Microbalance with Dissipation monitoring (E-QCM-D), Electrochemical Optical Waveguide Lightmode Spectroscopy (EC-OWLS) and Electrochemical Surface Plasmon Resonance (E-SPR), with the electro-photonic technology shown here (Fig. 4.34).

E-QCM-D has been proven to be a valuable tool for studying molecular and biomolecular interaction in parallel with electrochemical monitoring, with high detection sensitivity [3]. A unique feature of QCMD-based approaches is the compatibility with a wide range of surface materials, as the need to consider special properties such as optical transparency or reflectivity is not present (as long as the selected material can be deposited as a thin film onto the sensor crystals). In contrast to optical tech-

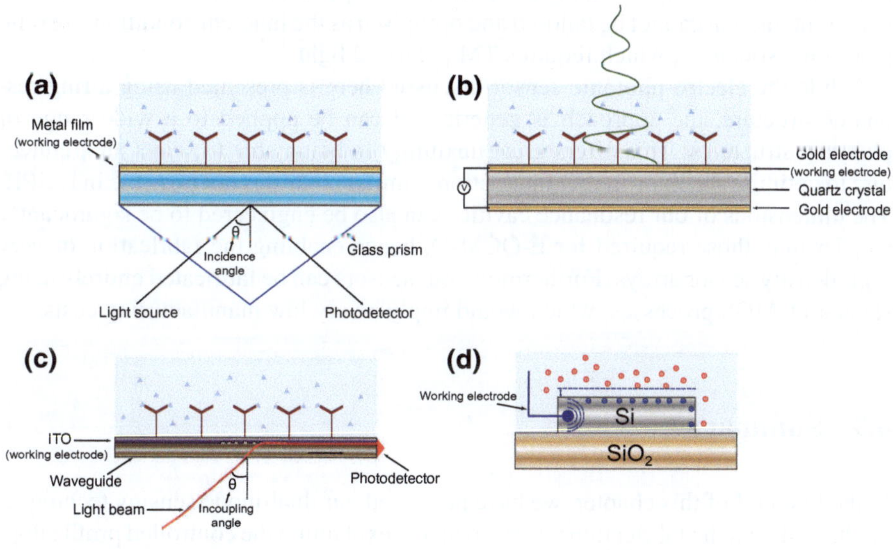

Fig. 4.34 Illustration of the sensing mechanisms of similar bi-domain approaches. **a** Electrochemical Surface Plasmon Resonance (E-SPR). **b** Electrochemical Quartz crystal Microbalance with Dissipation monitoring (E-QCM-D). **c** Electrochemical Optical Waveguide Lightmode Spectroscopy (EC-OWLS). **d** Electro-photonic silicon biosensing

niques, QCMD it is very sensitive to water associated with adsorbed proteins [44], but this can present a drawback in some situations as coupled water cannot be deconvoluted easily from the measurement. In terms of large-scale integration however, E-QCMD it is more limited than optical techniques, as the resonance wavelength of the quartz crystals is larger than those used in optical techniques, and therefore the theoretical minimum physical sensor size required to obtain a high Q-factor resonator is bigger [3].

In an OWLS system with electrochemical control (EC-OWLS), both TE and TM modes can be excited, enabling information about both the refractive index and thickness of the absorbed film to be extracted [45]. This combination of information provided from both optical modes, together with its inherent high sensitivity, makes this technology very well suited for studies of protein absorption kinetics [46]. This sensitivity is slightly higher than that obtained with E-QCM-D and E-SPR. However, this platform is limited, as only highly transparent and electrochemically compatible surfaces can be investigated. In addition, it is further limited by the inherent reduction in sensitivity of the sensing mechanism, as the evanescence tail of the confined optical mode propagates through the ITO film.

E-SPR exhibits high sensitivity towards local changes in the refractive index, enabling the detection of changes in the immobilised mass of molecules with a sensitivity comparable to E-QCM-D [47]. This sensing can be carried out with other functional surfaces when deposited on top of the metal surface. In contrast to E-QCM-D and E-OWLS, E-SPR can also be used to create 2D images of molecular interactions. This imaging is possible as plasmon resonances can be highly localised, however the sensitivity is slightly lower in this configuration [48]. However, the light-matter interaction cannot be tailored and optimised as the inherent condition to excite plasmon resonances, which requires TM polarised light.

While the electro-photonic sensor discussed here is presented using a ring resonator structure, the approach is generic and can be applied to a wide range of photonic structures. This offers great flexibility in being able to select an architecture to optimise the light-matter interaction. Similar control is not possible in E-SPR. The dimensions of our resonance cavities can also be engineered to be significantly smaller than those required for E-QCM-D, hence enabling the fabrication of very high-density sensor arrays. Furthermore, our sensors can be fabricated entirely using standard CMOS processes, which would imply a very low manufacturing costs.

4.9 Summary

In the first half of this chapter, we have presented our dual-mode sensing technique. We have shown the fabrications steps involved, explaining the controlled profile doping technique and the optimisation of the high-Q ring resonator cavity. The profile of the doping was characterised in depth to demonstrate its sharp gradient, which allows us to obtain resonances with a Q-factor as high as 50,000. An assessment of the correlation between the doping and the Q-factor was also provided, which models our system extremely well. In the second half of this chapter, the device was tested and an example of dual-mode sensing was provided. The marginal impact of doping on the device performance was demonstrated, as our device shows an *LOD* similar to other ring resonator based sensors reported in the literature. Subsequently, a layer of electro-active molecules was bound to the surface and insight into its (bio)chemical activity was presented. We show how it is possible to accurately extract the molecular density of immobilised layers of molecules on the surface, a measurement that is not possible with photonic detection alone unless making poorly defined assumptions about the properties of the layer. Finally, the limitations of the technique were mentioned and we compared it to other bi-domain sensing technologies.

References

1. J.A. Hubbell, J.P. Bearinger, J. Vrs, M. Textor, Electrochemical optical waveguide lightmode spectroscopy (ec-owls): A pilot study using evanescent-field optical sensing under voltage control to monitor polycationic polymer adsorption onto indium tin oxide (ito)-coated waveguide chips. Biotechnol. Bioeng. **82**(4), 465–473 (2003)
2. S. Xiaonan, K.J. Foley, S. Wang, X. Huang, N. Tao, Electrochemical surface plasmon resonance: Basic formalism and experimental validation. Anal. Chem. **82**(3), 935–941 (2010)
3. A. Krozer P. Brzezinski M. Rodahl, F. Hk, B. Kasemo. Quartz crystal microbalance setup for frequency and qfactor measurements in gaseous and liquid environments. Rev. Sci. Instrum. 3924–3930 (1995)
4. K.N. Rao, Optical and electrical properties of indium-tin oxide films. Pure Appl. Phys. **42**, 201–204 (2004)
5. N. Puri, V. Sharma, V.K. Tanwar, N. Singh, A.M. Biradar, Rajesh, Enzyme-modified indium tin oxide microelectrode array-based electrochemical uric acid biosensor. Progress Biomater. **2**(1), 1–7 (2013)
6. C.K. Choi, C.H. Margraves, S.I. Jun, A.E. English, P.D. Rack, K.D. Kihm, Opto-electric cellular biosensor using optically transparent indium tin oxide (ito) electrodes. Sensors **8**(5), 3257 (2008)
7. P.R. Solanki, A. Kaushik, V.V. Agrawal, B.D. Malhotra, Nanostructured metal oxide-based biosensors. NPG Asia Mater. **3**, 17–24 (2011)
8. M. Ibanescu, P. Bermel, J.D. Joannopoulos, A.F. Oskooi, D. Roundy, S.G. Johnson, Meep: a flexible free-software package for electromagnetic simulations by the fdtd method. Comput. Phys. Commun. **181**, 687–702 (2010)
9. Y.A. Vlasov, S. Assefa, C. Kang, C.T. Phare, S.M. Weiss, Photonic crystal slab sensor with enhanced surface area. Opt. Express **18**, 27930–27937 (2010b)
10. M.G. Scullion, T.F. Krauss, A. Di Falco, Slotted photonic crystals for biosensing applications. Proc. SPIE **8425**, 842507–842508 (2012)
11. N. Descharmes, Ulagalandha Perumal Dharanipathy, Zhaolu Diao, Mario Tonin, and Romuald Houdré. Observation of backaction and self-induced trapping in a planar hollow photonic crystal cavity. Phys. Rev. Lett. **110**, 123601–123605 (2013)
12. K. Rajeshwar. *Fundamentals of Semiconductor Electrochemistry and Photoelectrochemistry*, 1st edn. (Wiley, 2007)
13. A.C. Barrios, M.J. Ba nuls, V. González-Pedro, K.B. Gylfason, B. Sánchez, A. Griol, A. Maquieira, H. Sohlström, M. Holgado, R. Casquel. Label-free optical biosensing with slot-waveguides. Opt. Lett. **33**(7), 708–710 (2008)
14. R.N. Thomas, D.K. Schroder, J.C. Swartz, Free carrier absorption in silicon. IEEE J. Solid-State Circuits **13**(1), 180–187 (1978)

15. J. Isenberg, W. Warta, Free carrier absorption in heavily doped silicon layers. Appl. Phys. Lett. **84**(13), 2265–2267 (2004)
16. E.F. Schubert. *Doping in III-V Semiconductors* (Cambridge University Press, 1993)
17. D. Mathiot, J.C. Pfister, Dopant diffusion in silicon: a consistent view involving nonequilibrium defects. J. Appl. Phys. **55**, 3518–3530 (1984)
18. D. Klaassen, A unified mobility model for device simulationi. model equations and concentration dependence. Solid State Electron. **7**, 953–959 (1992a)
19. D. Klaassen, A unified mobility model for device simulationii. temperature dependence of carrier mobility and lifetime. Solid State Electron. **7**, 961–967 (1992b)
20. F.M. Smits, Measurement of sheet resistivities with the four-point probe. Bell Syst. Techn. J. **34**, 711–718 (1958)
21. D. Schroder, *Semiconductor Material and Device Characterization*, 3rd edn. (IEEE Press, Wiley, 2006)
22. Ltd. Shanghai simgui technology Co. (1996), http://www.simgui.com.cn/en/
23. K.L. Chopra, S.R. Das, *Thin Film Solar Cells*, 1st edn. (Springer, US, 1983)
24. W. Schottky, Zur halbleitertheorie der sperrschicht- und spitzengleichrichter. Zeitschrift fr Physik **113**, 367–414 (1939)
25. S.M. Sze, K.K. Ng. *Metal-Semiconductor Contacts*. (John Wiley and Sons, Inc., 2006)
26. B.J. Van Zeghbroeck (1996), http://ecee.colorado.edu/~bart/book/extrinsi.htm
27. H.C. Card, Silicon schottky barriers and ohmic contacts in integrated circuits. IEEE Trans. Electron Devices **23**(6), 538–544 (1976)
28. E. Philofsky, Intermetallic formation in gold-aluminum systems. Solid State Electron. **13**(10), 1391–1394 (1970)
29. E.H. Hall, On a new action of the magnet on electric currents. Am. J. Math. **2**, 287–292 (1879)
30. LJ van der PAUYV, A method of measuring specific resistivity and hall effect of discs of arbitrary shape. Philips Res. Rep. **13**, 1–9 (1958)
31. W. Bogaerts, P. De Heyn, T. Van Vaerenbergh, K. De Vos, S. Kumar, T. Selvaraja, P. Claes, P.B. Dumon, D. Van Thourhout, R. Baets, R. Baets, Silicon microring resonators. Laser Photonics Rev. **6**(1), 47–73 (2012)
32. X. Fan, Y. Sun, Optical ring resonators for biochemical and chemical sensing. Anal. Bioanal. Chem. **399**, 205–211 (2011)
33. M. Iqbal et al., Label-free biosensor arrays based on silicon ring resonators and high-speed optical scanning instrumentation. EEE J. Sel. Topics Quantum Electron. **16**, 654–661 (2010)
34. J. Niehusmann et al., Ultrahigh-quality-factor silicon-on-insulator microring resonator. Opt. Express **29**, 2861–2863 (2004)
35. K. De Vos, Label-free silicon photonics biosensor platform with microring resonators. Ph.D thesis, University of Ghent, 2010
36. Genalyte. Genalyte (2009), http://www.http://genalyte.com/
37. R. Soref, M. Nedeljkovic, G.Z. Mashanovich, Free-carrier electrorefraction and electroabsorption modulation predictions for silicon over the 1–14um infrarred wavelength range. IEEE Photon. J. **3**, 1171–1180 (2011)
38. L. Pukenas, R. Azhar, M. Guillorit, C. Walti, K. Critchley, S. Johnson, J. Murray, D. Nowak, R.S. Bon, Solid phase synthesis of functionalised sam-forming alkanethiol oligoethyleneglycols. J. Mater. Chem. B **2**, 3741–3744 (2014)
39. S. Johnson, D. Evans, S. Laurenson, D. Paul, A.G. Davies, P. Ko Ferrigno, C. Wlti, Surface-immobilized peptide aptamers as probe molecules for protein detection. Anal. Chem. **80**, 978–983 (2008)
40. I. Teraoka, S. Arnold, Estimation of surface density of molecules adsorbed on a whispering gallery mode resonator: utility of isotropic polarizability. J. Appl. Phys.**102** (2007)
41. D. Langford, R. Bonand, E. Koutsoumpeli, J. Murray, S. Johnson, Probing molecular interactions with methylene blue derivatized self-assembled monolayers. Sens. BioSens. Res. 1–6 (2015)
42. R.A. Soref, B.R. Bennett, Electrooptical effects in silicon. IEEE J. Quant. Electron. **23**, 123–129 (1987)

43. R.R. Gagne, C.A. Koval, G.C. Lisensky, Ferrocene as an internal standard for electrochemical measurements. Inorg. Chem. **19**(9), 2854–2855 (1980)
44. D.N. Furlong, F. Caruso, P. Kingshott, Characterization of ferritin adsorption onto gold. J. Colloid Interface Sci. **186**(1), 129–140 (1997)
45. J.J. Ramsden, Review of new experimental techniques for investigating random sequential adsorption. J. Stat. Phys. **73**(5), 853–877 (1993b)
46. R. Kurrat, B. Wlivaara, A. Marti, M. Textor, P. Tengvall, J.J. Ramsden, N.D. Spencer, Plasma protein adsorption on titanium: comparative in situ studies using optical waveguide lightmode spectroscopy and ellipsometry. Colloids Surf. B Biointerfaces **11**(4), 187–201 (1998)
47. F.R.M. Nirschl, J. Vrs, Review of transducer principles for label-free biomolecular interaction analysis. Biosensors **1**, 70–92 (2011)
48. J. Homola, Surface plasmon resonance sensors for detection of chemical and biological species. Chem. Rev. **108**(2), 462–493 (2008)

Chapter 5
Study and Application of Electrografted Layers of Diazonium Ions

Here we describe how the inclusion of electrochemical characterization methods alongside photonic sensing not only allows us to combine the complementary information contained within the two measurement domains, but also to exploit electrochemical processes to selectively modify the silicon surface and to regulate the local surface chemistry. Electrografting of three diazonium ions is studied and it is shown they provide specific chemical modification of the sensor surface. Furthermore, electrografting of diazonium ions enables site-selective control over the surface chemistry, hence paving a route towards multiplexed electro-photonic sensor arrays on the microscale.

5.1 Introduction

Our innovative sensor concept provides deeper insight into electrochemical reactions at the sensor surface than provided by electrochemical measurements alone. For example, here the formation of electrografted layers of aryl diazonium salts is studied for alkyne (CCH), maleimide ($H_2C_2(CO)_2NH$) and azide ($-N_3$) terminated diazonium salts. These diazonium salts provide specific surface modification that enable selective binding of biomacromolecules to the surface. Moreover, due to the fact that the electrografting reaction is spatially localised, diazonium ions can be selectively electrografted on different areas of a surface. We exploit this here to demonstrate a multiplexed electro-photonic sensor array in which each microring within the array is functionalised selectively with a different 'probe' molecule.

J. Juan Colaś, *Dual-Mode Electro-photonic Silicon Biosensors*, Springer Theses,
DOI 10.1007/978-3-319-60501-2_5

5.2 Application of Diazonium Ions

Aryl diazonium salts have been used extensively in organic chemistry due to the capability of introducing a wide range of functional groups into an aromatic ring to form an extensive variety of compounds [1]. They consist of an organic group (R in Fig. 5.1) sharing an aryl functional group with a dinitrogen cation (Fig. 5.1).

Since aryl diazonium salts were introduced [2], they have received considerable attention due to their ease of preparation, rapid (electro)reduction, large choice of reactive functional groups and strong arylsurface covalent bonding. Grafted layers of these compounds have been used in a large number of applications such as binding of macromolecules to materials of diverse nature and shapes and for controlling the hydrophobic/hydrophilic character of a surface [3]. Furthermore, diazonium salts are excellent alternatives to the traditional silanes (Sect. 2.5.2) used for surface functionalisation. Indeed, they can be easily reduced on carbon, metal or semiconductor surfaces [1]. The electrografting mechanism is illustrated in Fig. 5.2.

The electrografting of aryl layers is a straightforward procedure: when a diazonium salt is dissolved in an aprotic medium (e.g. a medium that does not contain hydrogen ions) with a supporting electrolyte, the application of a sufficiently large reductive potential will cleave the dinitrogen ion and so the aromatic ring is covalently attached to the surface. Figure 5.3 shows a typical cyclic voltammogram trace that characterises the electrografting of an aryl diazonium salt on a doped silicon electrode ($N_D = 7.5 \times 10^{16}\,\text{cm}^{-3}$) (Sect. 3.3.3).

In the first cycle of the CV measurement (i.e. first voltammogram) (Fig. 5.2), a prominent and irreversible reduction peak is observed at -1 V. The fact that the reduction peak is irreversible indicates that an irreversible electrochemical reaction has occurred associated with the cleavage of the dinitrogen ion. It is at this point when

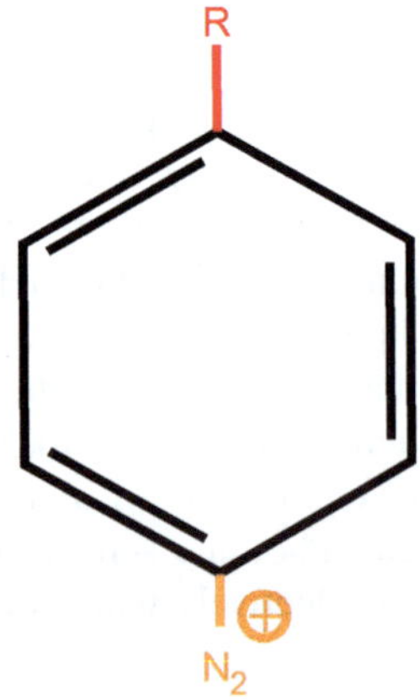

Fig. 5.1 General chemical structure of aryl diazonium salts. R (*red colour*) stands for a variety of functional groups, which can be organic or organometallic [1]

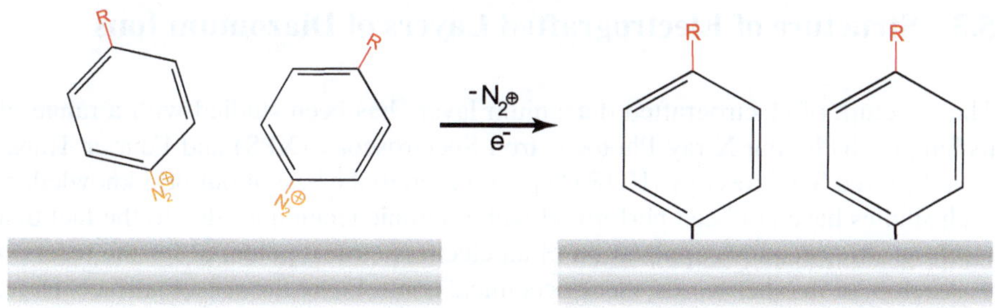

Fig. 5.2 Electrochemical grafting of a generic diazonium salt to a surface. When applying a reductive voltage, the dinitrogen ion is cleaved leading to covalent attachment of the aromatic ring to the electrode surface

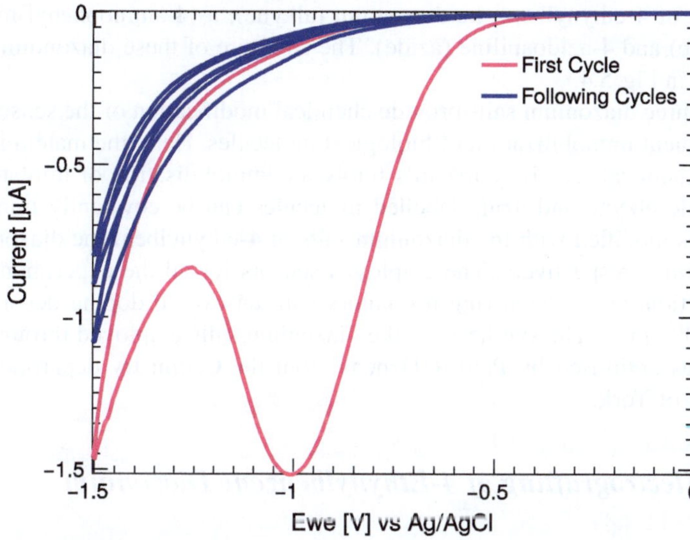

Fig. 5.3 Typical electrografting of an aryl diazonium salt. The magenta trace represents the first cycle of the CV measurement where a broad and irreversible reduction peak at −1 V represents the cleavage of the dinitrogen ion and the grafting of the molecule to the surface. In subsequent voltammograms (*blue traces*), no additional peaks are observed as the surface is completely covered by the electrografted molecules, restricting the electrografting of further molecules

the organic groups become covalently bonded to the surface. On following cycles, no additional reduction peaks are observed as the surface is completely covered by the electrografted molecules, preventing the electrografting of further molecules.

The high grafting density of aryl groups leads to the formation of well-ordered and homogenous monolayers of controlled structure. The density of the electrografted layer can be calculated directly from the cyclic voltammogram as described in Sect. 4.6. For electrografted aryl diazonium layers, molecular surface densities in the range of $10^{13} - 10^{14}$ molecules/cm^2 are expected [3], which is similar to the density of SAMs of alkanthiols assembled on gold [4].

5.3 Structure of Electrografted Layers of Diazonium Ions

The structure of electrografted diazonium layers has been studied with a range of techniques including X-ray Photoelectron Spectroscopy (XPS) and Fourier Transform Infrared Spectroscopy (FTIR) [3]. However, to the best of our best knowledge, such studies have not been performed with photonic techniques due to the fact that electrografting requires application of an electrochemical potential for the reaction to develop, or in other words, electrochemical control over the sensor surface. Here, given the novelty of our sensor (the combination of electrochemical and optical information), we have been able to explore the structure of such electrografted layers in both electrochemical and optical domains, providing deeper insight than obtained by electrochemical measurements alone. Here, we study in situ generated diazonium salts of 4-ethynylbenzene diazonium (alkyne), n-(4-aminophenyl)maleimide (maleimide) and 4-azidoaniline (azide). The structure of these diazonium salts are illustrated in Fig. 5.4.

These three diazonium salts provide chemical modification of the sensor surface for subsequent immobilization of biological molecules. Here, the maleimide group provides chemical reactivity towards thiols for immobilisation of thiolated molecules, while alkyne and azide labelled molecules can be covalently immobilised on surfaces modified with the diazonium salts of 4-ethynylbenzene diazonium and 4-azidoaniline, respectively. The employed sensors for all the experiments shown in this section were silicon ring resonators with an overall doping density of $N_D = 7.5 \times 10^{16}\,\mathrm{cm}^{-3}$. The synthesis of the diazonium salts employed throughout this section was optimised by Patrick Deneny from the Chemistry department of the University of York.

5.3.1 *Electrografting of 4-Ethynylbenzene Diazonium*

For the generation of the molecule 4-ethynylbenzene diazonium, we used the molecule 4-ethynylaniline (Sigma-Aldrich). The preparation method compromises the

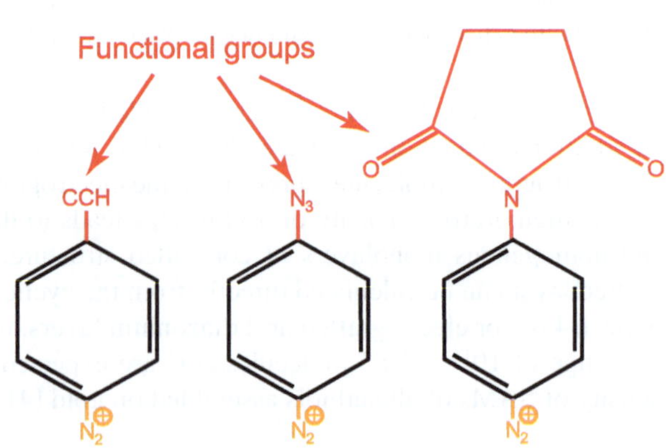

Fig. 5.4 Structure of the aryl diazonium salts studied here. Alkyne (*left*), azide (*centre*) and maleimide (*right*) were the organic functional groups (*red*)

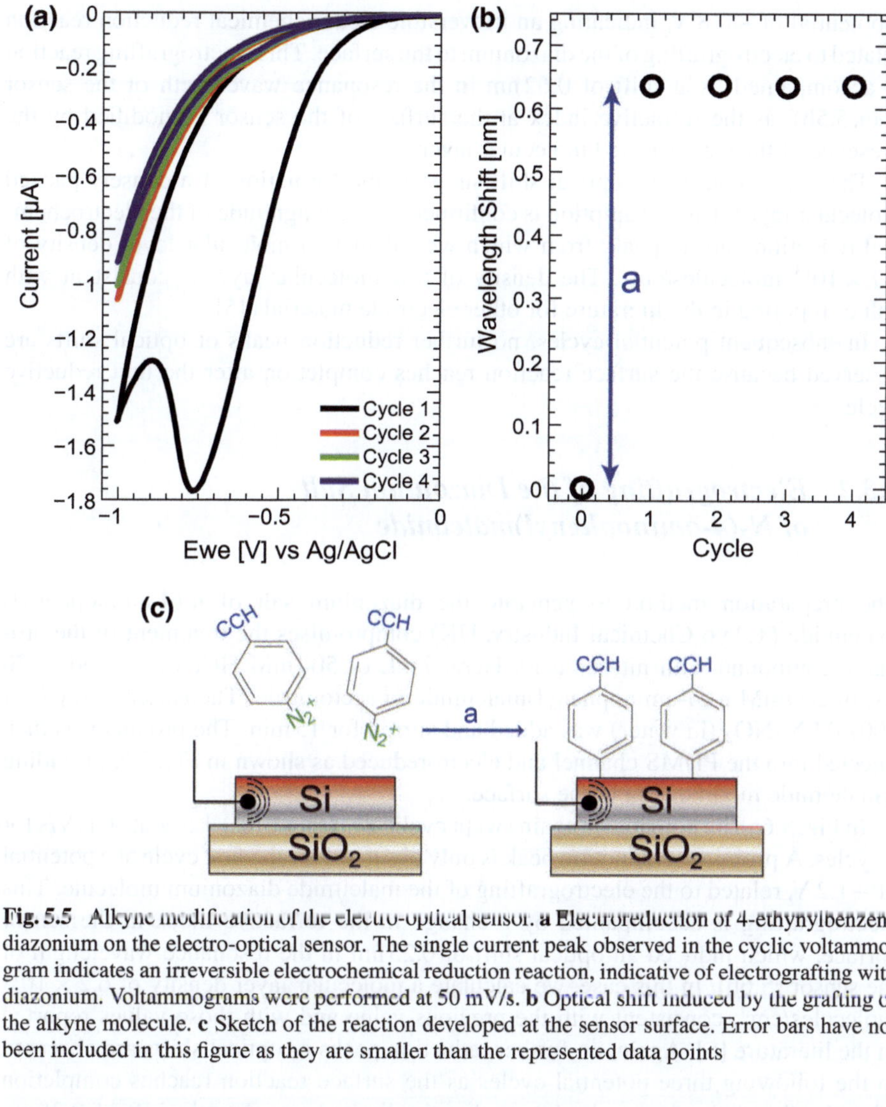

Fig. 5.5 Alkyne modification of the electro-optical sensor. **a** Electroreduction of 4-ethynylbenzene diazonium on the electro-optical sensor. The single current peak observed in the cyclic voltammogram indicates an irreversible electrochemical reduction reaction, indicative of electrografting with diazonium. Voltammograms were performed at 50 mV/s. **b** Optical shift induced by the grafting of the alkyne molecule. **c** Sketch of the reaction developed at the sensor surface. Error bars have not been included in this figure as they are smaller than the represented data points

treatment of the aromatic compound with nitrous acid. The nitrous acid was generated in situ from a mixture of hydrochloric acid (HCl, Fisher Scientific) and sodium nitrite ($NaNO_2$, Fisher Scientific). 35 mg of 4-ethynylaniline was dissolved in 7 mL of 500 mM HCl at room temperature. Then 900 μL of 1 M aqueous $NaNO_2$ was slowly added to the mixture and left for 10 min with continuous stirring. The product, which provides alkyne modification to the surface, was then injected into the PDMS channel and electroreduced as shown in Fig. 5.5.

In Fig. 5.5a, the electrode potential is swept cyclically at 50 mV/s from 0 to −1 V for 4 cycles. A prominent reduction peak is observed for the first cycle that occurs at a potential of −0.8 V, indicating an irreversible electrochemical reduction reaction related to electrografting of the diazonium to the surface. This electrografting reaction is accompanied by a shift of 0.62 nm in the resonance wavelength of the sensor (Fig. 5.5b), as the refractive index at the surface of the sensor is modified by the presence of the eletrografted molecular layer.

The magnitude of the optical shift suggests the formation of a densely packed molecular layer. This assumption is confirmed by the magnitude of the electrochemical reduction current peak, from which we calculate a molecular layer density of 3.7×10^{13} molecules/cm^2. The density of this molecular layer is consistent with values reported in the literature for other electrode materials [5].

In subsequent potential cycles, no further reduction peaks or optical shifts are observed because the surface reaction reaches completion after the first reductive cycle.

5.3.2 Electrografting of the Diazonium Salt of N-(4-aminophenyl)maleimide

The preparation method to generate the diazonium salt of n-(4-aminophenyl) maleimide (Tokyo Chemical Industry, UK) compromises the treatment of the aromatic compound with nitrous acid. Here, 7 mL of 500 mM HCl was added to 70 μL of 200 mM n-(4-aminophenyl)maleimide in acetonitrile. Thereafter, 280 μL of 100 mM $NaNO_2$ (in water) was added and stirred for 15 min. The product was then injected into the PDMS channel and electroreduced as shown in Fig. 5.6, providing a maleimide modification to the surface.

In Fig. 5.6a, the potential is again swept cyclically from 0 to −1.5 V at 50 mV/s for 4 cycles. A prominent reduction peak is only observed for the first cycle at a potential of −1.2 V, related to the electrografting of the maleimide diazonium molecule. This electrografting is accompanied by a change in the refractive index at the sensor surface, which induced an optical shift of 0.25 nm in the resonance wavelength of the sensor (5.6b). In this case, we calculate a molecular layer density of 6.2×10^{13} molecules/cm^2, consistent with the previous value and with those values reported in the literature [5]. Again, no further reduction peaks or optical shifts are observed in the following three potential cycles as the surface reaction reaches completion after the first reductive cycle. I note that the limited wavelength shift of 0.25 nm obtained in this experiment was obtained with a ring resonator device which had lower sensitivity than those used in Sects. 5.3.1 and 5.3.3, as its cavity waveguide was wider.

5.3.3 Electrografting of 4-Azidoaniline

For the synthesis of the third in situ generated diazonium salt employed here, the preparation method follows the same guidelines. The molecule 4-azidoaniline (Sigma-Aldrich) is used in this case, which is again treated with nitrous acid. 7 mL of

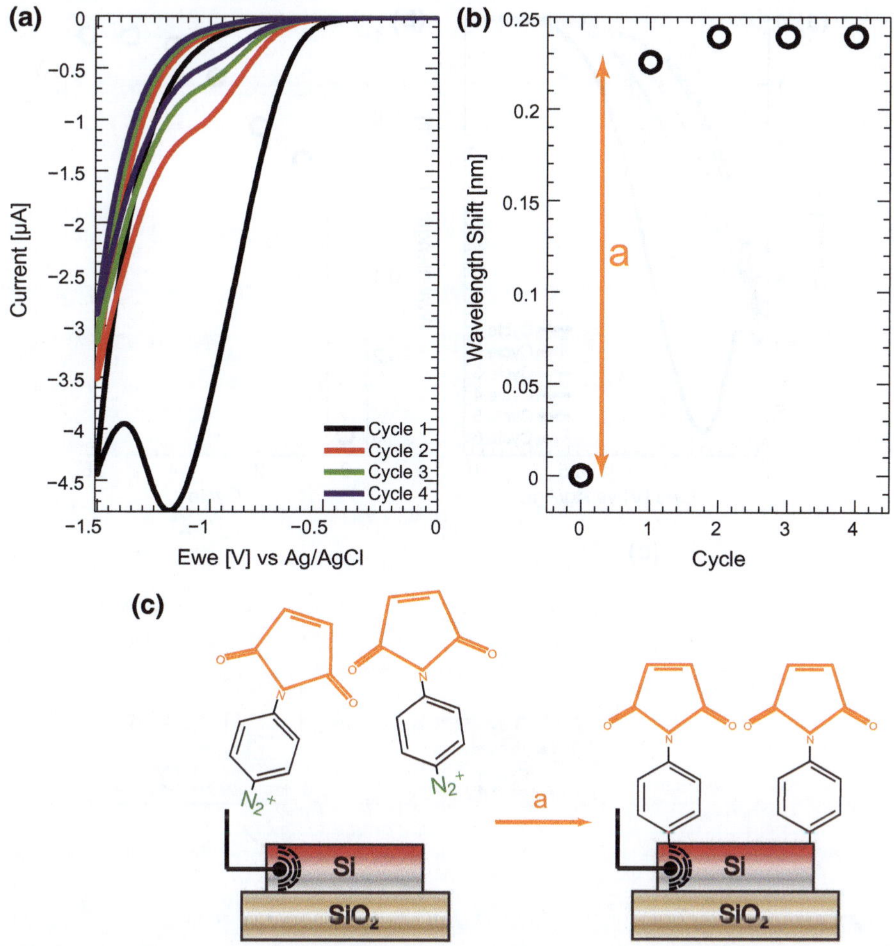

Fig. 5.6 Maleimide modification of the electro-optical sensor. **a** Electroreduction of the diazonium salt of n-(4-aminophenyl)maleimide on the electro-optical sensor. A single current peak, indicative of electrografting with diazonium, is observed in the first cyclic voltammogram (performed at 50 mV/s). **b** Optical shift induced by the grafting of the maleimide molecule. **c** Sketch of the reaction developed at the sensor surface. For simplicity, the chemical formula $H_2C_2(CO)_2NH$ has been replaced with the label 'Maleimide'. Error bars have not been included in this figure as they are smaller than the represented data points

500 mM HCl was added to 70 μL of 100 mM 4-azidoaniline in methanol, then 280 μL of 100 mM $NaNO_2$ was added and stirred for 15 min. The product, injected into the microfluidic channel, was then electroreduced as shown in Fig. 5.7, providing an azide modification to the sensor surface.

As shown in Fig. 5.7, the case of 4-azidoaniline is different to those of 4-ethynylbenzene diazonium and n-(4-aminophenyl)maleimide. When 4-azidoaniline is electrografted (Fig. 5.7a), each of the first four voltammogram cycles contain reductive

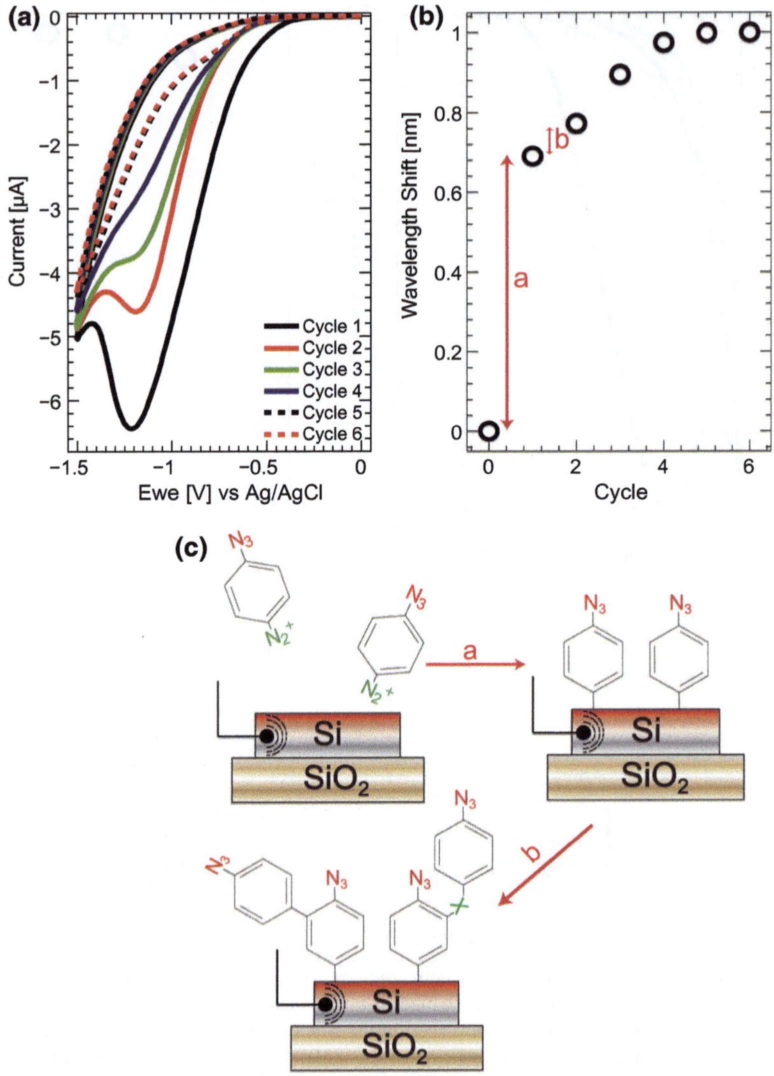

Fig. 5.7 Azide modification of the electro-optical sensor. **a** Electroreduction of the diazonium salt of 4-azidoaniline on the electro-optical sensor leads to multiple peaks in the reduction current that occur with each 50 mV/s cycle of the electrode potential. **b** The photonic response provides insight into this reaction, as wavelength shifts associated with each cycle are also observed. **c** Sketch of the reaction developed at the sensor surface. X indicates an unspecified chemical functionality. Error bars have not been included in this figure as they are smaller than the represented data points

potential peaks of decreasing amplitude. Each of these potential cycles also leads to a corresponding shift in the resonance wavelength of our sensor (Fig. 5.7b).

Here, the combination of photonics and electrochemistry provides additional insight into the reactions, as the optical information confirms the electro-reduction processes measured in cycles 1–4 results from a modification of the sensor surface, rather than from an electrochemical reaction of excess diazonium ions in the solution-phase. The magnitude of the first reduction peak corresponds to a molecular surface density of 2.1×10^{14} molecules/cm^2. This molecular surface density, together with the large optical shift of 0.65 nm observed after the first potential cycle, indicates the formation of a complete monolayer on the silicon sensor surface. The combination of these two sensing modes thus enables us to deduce that subsequent electroreduction processes in cycles 2–4 correspond to the formation of a multilayer structure at the sensor surface.

Multilayers are often observed in diazonium electrografting [3], and the mechanism of their formation involves highly reactive aryl radicals that are generated during diazonium electrografting (Fig. 5.7c). Our observation that multilayer formation occurs for the diazonium ion of 4-azidoaniline electrografting, but not for diazoniums ions of 4-ethynylbenzene and n-(4-aminophenyl)maleimide, is consistent with previous studies which have shown that the propensity for multilayer formation is highly dependent on the structure of the diazonium [6]. However, it is highlighted that multilayers are obtained when repeating voltage cycles, whereas a well-ordered and densely packed monolayer (desired for biofunctionalisation of surfaces) is obtained with a single voltage cycle.

5.3.4 *Control Experiments of the Electroreduction of Diazonium Salts*

Control experiments for the diazonium salts of 4-ethynylaniline and azidoaniline were performed in order to verify that both electrochemical reduction peaks and optical wavelength shifts are due to the electrografting of the in situ generated molecules. Figure 5.8 shows control experiments of the electroreduction of both azidoaniline and 4-ethynylbenzene diazonium solutions without adding the diazonium molecule.

As expected, no reduction peaks are observed in any of the voltammograms because no electroactive molecule is present in the solution. All other chemicals were added to the mixture, and cyclic voltammograms were measured at 50 mV/s.

Confirmation of the surface chemistry for the diazonium salt of 4-azidoaniline was also obtained through XPS. Figure 5.9 shows the spectrum of a silicon substrate without biomodification.

We observe a carbon contamination peak at 285 eV. This contamination peak provides a reference for the comparison between the biomodified silicon sample, which is shown in Fig. 5.10.

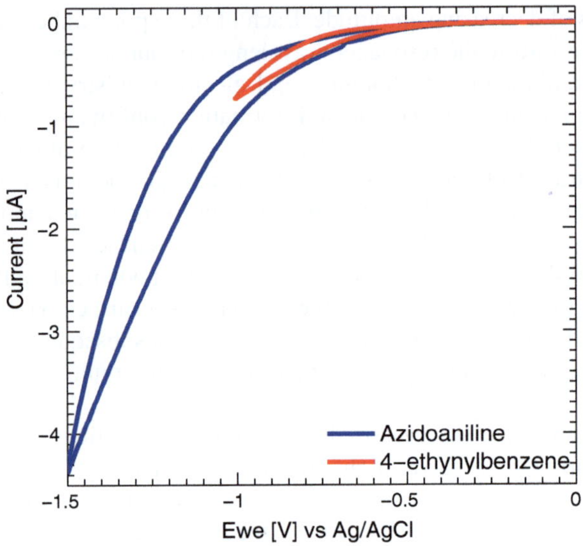

Fig. 5.8 Control experiments of the electroreduction of both azidoaniline and 4-ethynylbenzene diazonium solutions without adding the electroactive active molecule

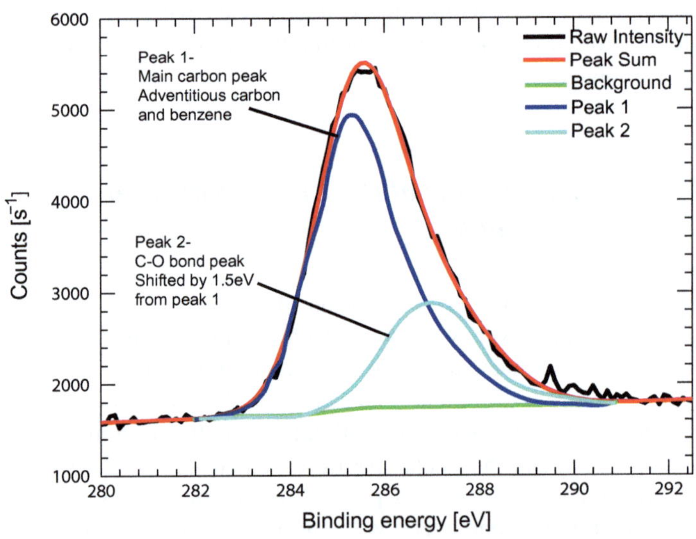

Fig. 5.9 X-ray photoelectron spectroscopy (XPS) spectrum of a silicon substrate without biomodification. The main carbon contamination peak is observed at 285 eV

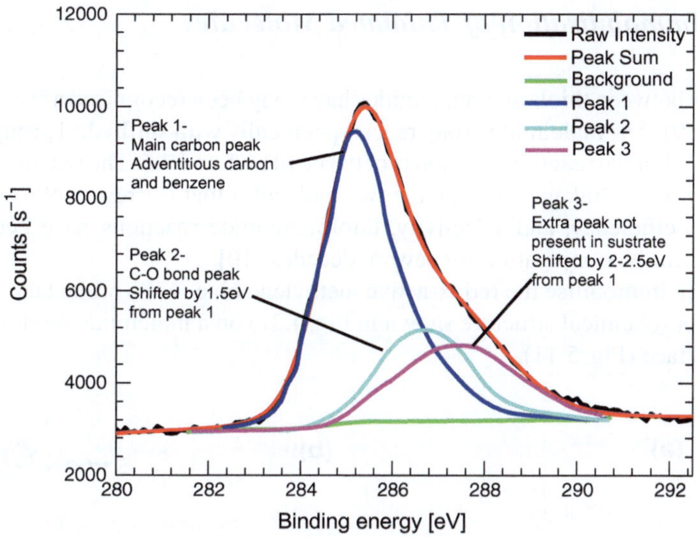

Fig. 5.10 XPS spectrum of a silicon substrate where the diazonium salt of 4-azidoaniline has been previously electrografted

Comparing the spectra of the unmodified silicon sample of Fig. 5.9 and the bio-modified sample with diazonium salt of 4-azidoaniline of Fig. 5.10, a new peak is observed. This new peak, shifted 2–2.5 eV from the main carbon peak, indicates the presence of the electrografted diazonium molecule on the surface. Furthermore, the main carbon peak (labelled Peak 1) is significantly greater following electroreduction, consistent with the formation of an organic molecular layer.

5.4 Functionalisation of Electrografted Layers of Diazonium Ions

Aryl diazonium chemistry offers a route to surface-biofunctionalisation. By synthesising a diazonium salt with the desired functional group (R in Fig. 5.1), functionalisation with, for example, receptor molecules for label-free sensing applications [7] can be obtained. Or also, immobilisation of electrochemical probes for electrochemical sensing [5].

Here we present two examples of surface-functionalisation using electrografted layers of diazonium salts. In the first one, the thiolated electro-ative molecule employed in Sect. 4.6 is immobilised on a maleimide terminated silicon electrode. In the second example, another electrochemical probe, namely ferrocene [8], is immobilised on an azide modified silicon surface through copper-catalysed azide-alkyne cycloaddition (CuAAC) click reaction.

5.4.1 Immobilisation of Thiolated Molecules

Reactions between thiols and maleimides have long been recognised as very efficient reactions [9]. The maleimide group reacts specifically with sulfhydryl groups (thiols) when the pH of the reaction solution is between pH 6.5 and 7.5. The reaction results in a stable thioester linkage between the two molecules that is irreversible. Given their reliability, efficiency, and selectivity, thiol-maleimide reactions have been widely employed for bioconjugation for several decades [10].

Here we immobilise the redox-active methylene blue (MB) probe labelled with a thiol-linker (chemical structure shown in Fig. 4.27) on a maleimide modified silicon sensor surface (Fig. 5.11).

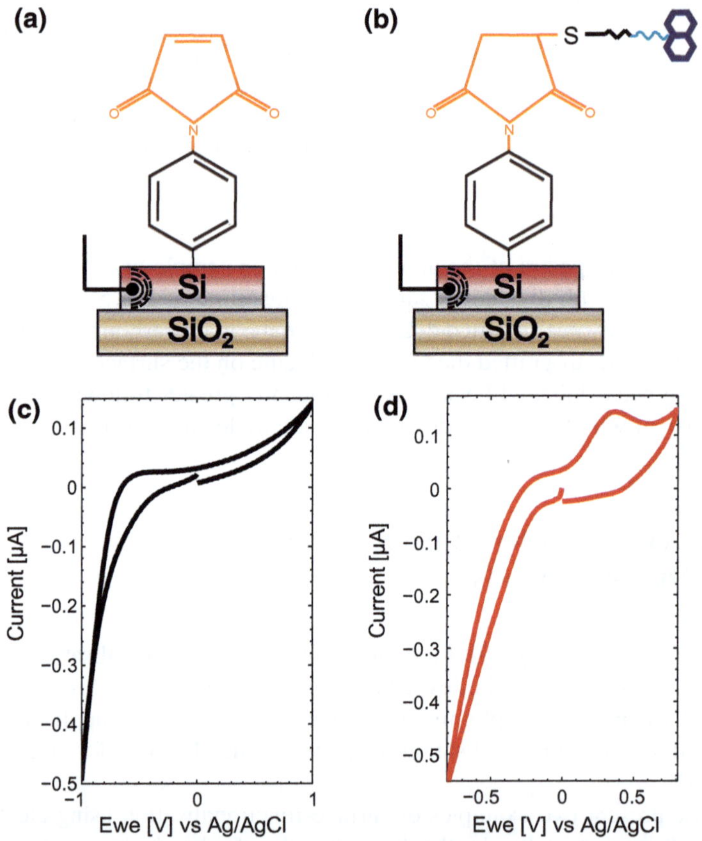

Fig. 5.11 Immobilisation of the redox-active MB probe labelled with a thiol-linker on a maleimide modified silicon substrate. **a** and **c** illustrate the surface modification, while the cyclic voltammograms in **b** and **d** show the electrochemical response of the maleimide and MB modified surface, respectively. Cyclic voltammetry was performed at 50 mV/s using a 100 mM potassium phosphate (PB) pH 7 electrolyte

Firstly, maleimide groups were electrgrafted onto the silicon substrate following the method described in Sect. 5.3.2. Once modified, as illustrated in Fig. 5.11a, the maleimide modified substrate displays no electrochemical activity (Fig. 5.11c). Subsequently, the maleimide modified surface is incubated with the MB probe at a concentration of 0.75 mM in ultrapure (MiliQ) water. After 60 min of incubation, the substrate is rinsed thoroughly leading to a surface functionalised with the MB probe (Fig. 5.11b) and electrochemical activity is observed using 100 mM PB pH 7 as electrolyte (Fig. 5.11d). An oxidation peak is observed at a voltage value of 0.35 V indicating the oxidation of the MB probe. We note the reduction peak is less prominent due to the large background current. From the area under the oxidation peak we calculate a surface density 2.56×10^{12} molecules/cm^2, which is similar to that reported on gold electrodes [11].

5.4.2 Functionalisation Through Copper-Catalyzed Azide-Alkyne Huisgen Cycloaddition (CuAAC)

Similarly to the maleimide-modified surface, azide-modified surfaces can be functionalised with molecules containing alkyne groups through the high yield CuAAC click reaction, and *vice versa* [12]. This CuAAC click reaction is used widely for the conjugation of molecular entities, and proceeds over a broad temperature range and is insensitive to aqueous conditions in a pH range between 4 and 12 [12]. Here, we used this click reaction to immobilise an alkyne modified version of the redox active probe Ferrocene [8] ($C_{10}H_{10}Fe$, Sigma Aldrich) to a azide modified Si surface.

To trigger the CuAAC process, copper ions in the +1 oxidation state are needed. These ions are most conveniently supplied by a mixture of a Cu^{2+} salt and a solution based reducing agent, such that Cu^{2+} ions are reduced to Cu^{1+}. Alternatively, the generation of these ions may be performed by in situ electroreduction of Cu^{2+} to Cu^{1+} [13]. Here, we generate Cu^{1+} ions using sodium ascorbate as the reducing agent.

Initially, the surface was azide modified by electrochemical reduction of in situ generated 4-azidodiazonium cations following the procedure described in Sect. 5.3.3. Once modified, 217 μL of 115 mM aqueous $CuSO_4$, 200 μL of 1 M ascorbic acid dissolved in ultrapure (MiliQ) water, 333 μL of 15 mM ethynylferrocene in ethanol, 1.25 mL of 400 mM of potassium hexafluorophosphate (KPF_6) in water and 3 mL of ethanol were combined to generate the necessary reaction solution. This was flowed over the azide modified silicon surface (Fig. 5.12a, b) and left overnight (in the dark) to enable development of the azide-alkyne Huisgen cycloaddition. Following this, the surface was thoroughly rinsed with water and ethanol to wash unreacted molecules from the surface. The electrochemical measurements were performed in 100 mM KPF_6 to characterise the functionalised surface as shown in Fig. 5.12c.

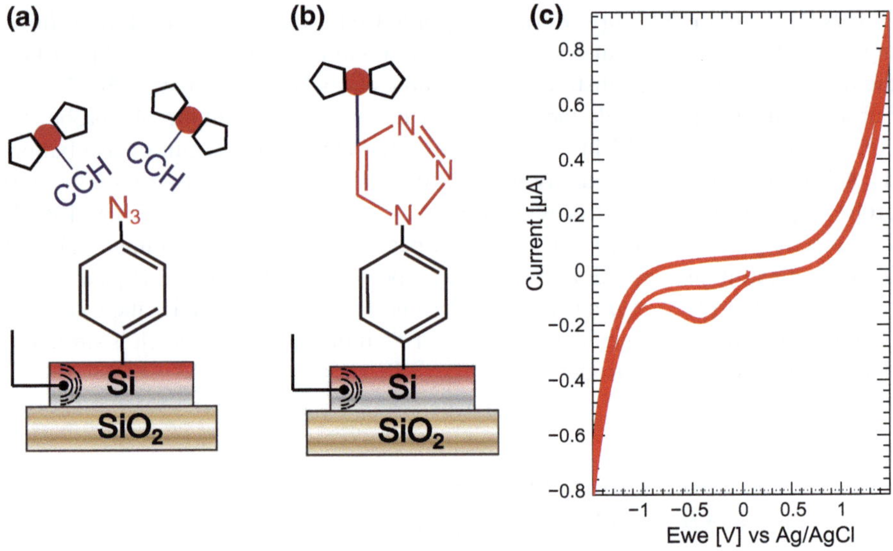

Fig. 5.12 Immobilisation of Ferrocene on an azide modified silicon substrate. **a** and **b** illustrate the CuAAC process at the surface of the substrate before and after cycloaddition, respectively. **c** shows the electrochemical response of the ferrocene functionalised silicon substrate

Measurements shown in Fig. 5.12c, display a reductive peak at a potential of −0.48 V, which corresponds to the reduction of the immobilised ferrocene molecule. The oxidation peak cannot be observed as the background current is higher than that induced by the oxidation of the molecule.

5.5 Site-Selective Functionalisation of Optical Biosensors

With the emergence of personalised medicine, accurate and specific detection of biologically and clinically relevant biomarkers in their native conformation will be required [14]. Such detection needs to be performed in parallel, which demands complex arrays of photonic sensors. To satisfy this need, high spatial resolution functionalisation is needed to accurately functionalise highly dense arrays of sensors.

In the literature, different selective functionalisation techniques have been described including microchannels for sample delivery [15–19], using site-specific antibody immobilisation strategies [20], inkjet printing [21] and lithography methods [22]. Nevertheless, such functionalisation techniques do not allow the development of highly dense photonic arrays due to their limited spatial resolution, which is on the order of hundreds of micrometers.

As electrografting is underpinned by electron tunneling [23], the reaction is highly spatially localised and occurs only at the electrode. This enables surface modification with high spatial resolution. For example, a resolution of 15 nm has been demonstrated using this approach [24]. It is thus possible to exploit electrografting of diazonium ions for site-selective control the surface chemistry suitable to the selective functionalisation of high density arrays of photonic sensors. Examples of selectively controlled electroreduction on silicon surfaces can be found in [25, 26], where locally doped areas have been used for such selective control.

Here, we have developed a proof-of-concept multiplexed photonic sensor array, with which we have been able to selectively control (temporally and spatially) the immobilisation of receptor molecules on each of our sensors within a highly compact array at a submicron scale. This array is composed of two identical microring resonators (Sect. 5.5.1), each of which is functionalised with a different 'probe' molecule, here two single stranded DNA molecules of different sequences. We subsequently demonstrate multiplexed detection of the complementary DNA sequences (Sect. 5.5.2). In this case, due to setup limitations (one single optical line) the measurements have been performed sequentially, but performing them in parallel is possible with the appropriate two-line setup.

5.5.1 Sensor Microarray Fabrication

The microarray is made of two identical optical ring resonators separated by a distance of 250 μm (Fig. 5.13a) and fabricated using the process described in Sects. 4.2, 4.3 and 4.4. To achieve the desired selective addressability, both sensors have to be electrically isolated. We achieve this isolation by etching a trench in the silicon layer between the two sensors, thereby electrically isolating the two ring cavities (Fig. 5.13b, c). Two contacts pads were then placed at the edges of each section to electrically address each ring resonator individually.

To obtain the trench, the positive photoresist S1818 (Shipley) was spin coated at 4000 rpm and baked at 100 °C for 2 min to obtain a 300 nm thick layer. Using a photomask, the resist is exposed with ultraviolet (UV) light for 4 min. Once exposed, the resist is developed for 40 seconds in MF-319 (Rohm and Haas Electronic Materials LLC). Thereafter, the patterned sample is dry etched for 2 min with RIE using the same conditions detailed in Sect. 4.4. Finally, the remaining resist is removed by immersion in acetone, rinsed in IPA and a PDMS microchannel is then bonded onto the sample following the method described in Sect. 3.2.5.

5.5.2 Selectively Functionalised Photonic DNA Microarray

In molecular biology, gene expression profiling is used to create an overview of cellular function, aimed at, for example, evaluating whether cells are responding to a particular treatment or not. This process involves thousands of genes being simultaneously monitored to study the effects of certain treatments or diseases. This procedure

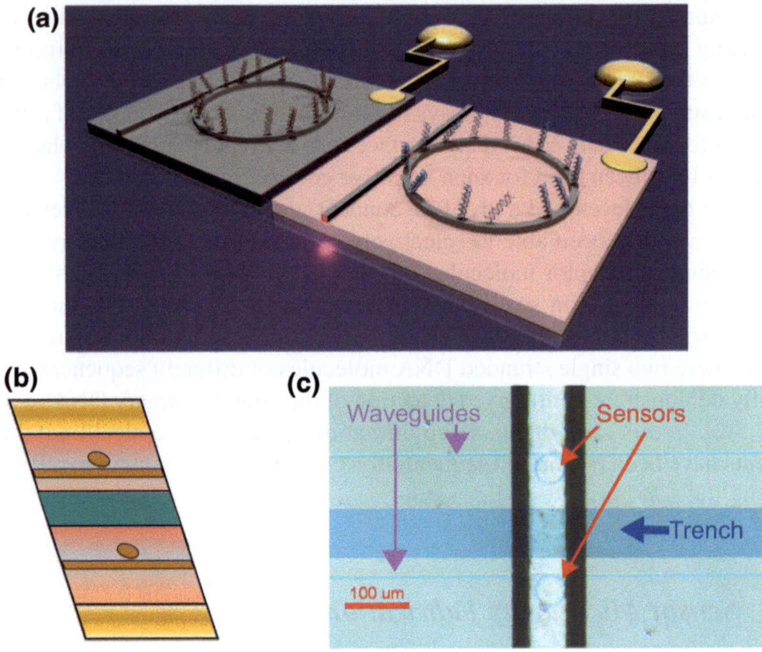

Fig. 5.13 Sensor microarray fabrication. **a** The sensor array is made of two identical optical ring resonators. **b** Two contact pads are placed at the edges of the sample to electrically address each ring resonator individually. A trench is etched between both sensors thus they are electrically isolated. **c** Optical microscope image of the final sensor array after a PDMS microchannel is bound

is performed by microarray-based devices, in which a collection of microscopic DNA spots (spot size typically around 50–100 μm) is attached to a solid surface [27].

A number of photonic-based approaches to fabricate such DNA microarrays can be found in the literature [15, 22, 28]. To date, however, no feasible approaches have been presented that are able to match the high density of electronic DNA microarrays [27].

Here, we demonstrate how our technology can provide a route for gene expression profiling using photonic sensors by demonstrating site specific DNA-bioconjugation at the micro-scale in a two ring resonator array.

Site-selectively functionalisation of an optical ring resonator array

Firstly, one sensor was modified with 4-ethynylaniline by electrografting following the method presented in Sect. 5.3.1. Then, the microfluidic channel was rinsed with ethanol to wash unreacted 4-ethynylaniline molecules. The diazonium salt of 4-azidoaniline was subsequently electrografted on the second sensor to provide an azide modification to this surface (using the method shown in Sect. 5.3.3). At this point, the surface of each individually addressable electro-photonic sensor was functionalised

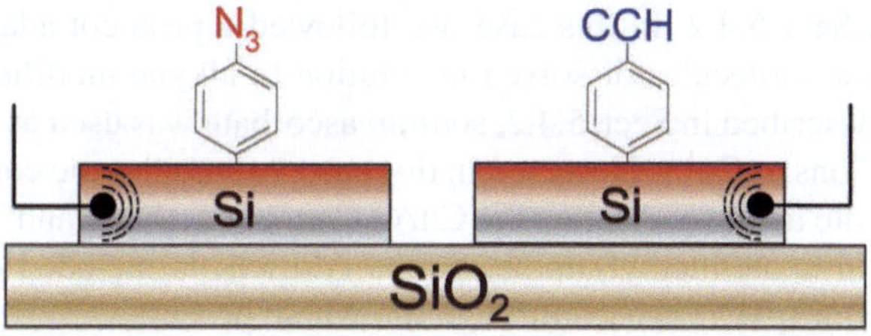

Fig. 5.14 Selectively functionalised optical ring resonator array. Each individually addressable electro-photonic sensors was functionalised with a different 'probe' to provide orthogonal chemical functionality

with a different 'probe' to provide orthogonal chemical functionality as depicted in Fig. 5.14.

DNA oligonucleotides design

Our photonic DNA microarray was designed to detect specific hybridisation of two different single-stranded DNA (ssDNA) oligonucleotides. The complementary sequences were immobilised selectively on the two ring resonators by the inclusion of alkyne or azide modification, as discussed in the following section.

The sequence with the azide modification and its reverse complement were randomly generated and then checked for secondary structure and unwanted interactions using the NUPACK software suite [29]. Similarly, the ssDNA strand with the alkyne modification and its reverse complement were designed to have no secondary structure and to be orthogonal to both the azide-modified strand and its reverse complement. These four ssDNA oligos are shown in table 5.1.

The melting temperature for the azide duplex is around 70–55 °C for the alkyne duplex (calculated using NUPACK considering the buffer 1xTris EDTA (1xTE) with 1 M NaCl). The strands were acquired from Integrated DNA Technologies (IDT).

Immobilisation of alkyne/azide modified DNA oligonucleotides

The two patterned regions within the array, with either an alkyne or azide modification, can link to azide or alkyne molecules, respectively, through the CuAAC reaction

Table 5.1 DNA oligonucleotides sequences used for the demonstration of the photonic DNA microarray

Label	Sequence	Notes
$ssDNA_{azide}$	5'-ACACGCATACACCCAT(3AzideN)-3'	Azide modified DNA
$cDNA_{azide}$	5'-ATGGGTGTATGCGTGT-3'	Reverse complement of the azide modified DNA
$ssDNA_{alkyne}$	5'-GTCATTTCTCTAAGTA(octadiynyl)-3'	Alkyne modified DNA
$cDNA_{alkyne}$	5'-TACTTAGAGAAATGAC-3'	Reverse complement of the alkyne modified DNA

as described in Sect. 5.4.2. In this case, we followed a protocol adapted from [30] which links azide molecules dissolved in solution to alkyne modified surfaces and *vice versa*. As described in Sect. 5.4.2, sodium ascorbate was used as reducing agent to reduce Cu^{2+} ions to Cu^{1+}. However, in this case, to stabilise the copper ions in the +1 oxidation state and to accelerate the CuAAC reaction, the ligand THPTA (tris(3-hydroxypro pyltriazolylmethyl)amine, Sigma Aldrich) is included in the mixture. 100 mM PB buffer (pH 7) is used through the whole process.

Initially, copper (II) sulfate ($CuSO_4$, Sigma Aldrich) was dissolved in ultrapure (MiliQ) water at a concentration of 20 mM. 250 μL were then added to a 500 μL solution of the ligand THPTA at 50 mM. The mixture was then added to either 2 μM alkyne or azide modified DNA in PB buffer (pH 7). Finally, 150 μL of sodium ascorbate at 1 mM concentration (in ultrapure water) was added to the solution.

The solution containing the alkyne-group DNA, $ssDNA_{alkyne}$, was injected through the microfluidic channel at a rate of 20 μL/min (the rate was maintained constant throughout the measurements). $ssDNA_{alkyne}$ can only react with the ring functionalised with azide groups as shown in Fig. 5.15a, b. The conjugation of $ssDNA_{alkyne}$ to the modified azide surface was monitored optically and in real-time as shown by the

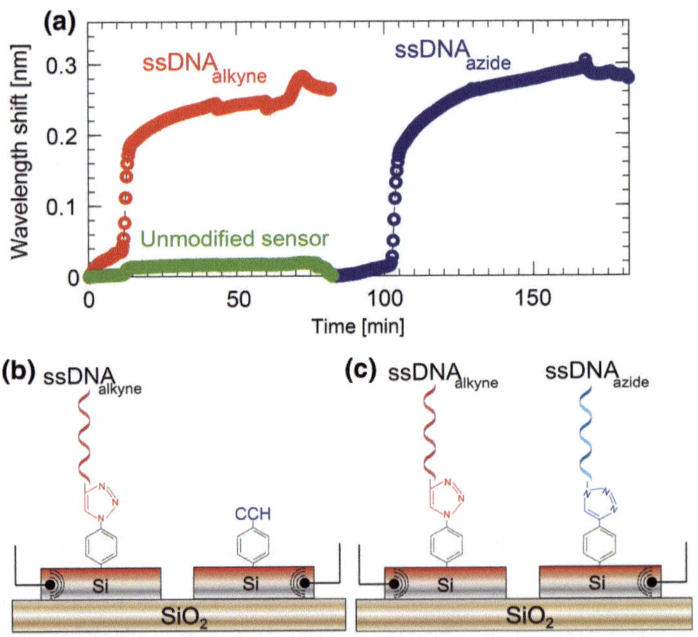

Fig. 5.15 Selective immobilisation of alkyne/azide modified DNA oligonucleotides. **a** Individual sensors are functionalised with modified DNA sequences using the CuAAC reaction, showing selective binding to the microrings functionalised with the pertinent chemical moiety. When no surface modification is present, no DNA immobilisation is observed as illustrated by the green trace. **b** Firstly the azide modified sensor is functionalised with the $ssDNA_{alkyne}$ sequence. **c** Secondly the alkyne modified sensor with the $ssDNA_{azide}$ sequence

red trace in Fig. 5.15a. Once the wavelength shift reaches saturation (0.25 nm shift after 60 min), the system was re-aligned to the second ring resonator and PB buffer (pH7) was injected to remove unreacted $ssDNA_{alkyne}$.

Secondly, the azide-group DNA, $ssDNA_{azide}$, was injected reacting just with the alkyne modified microring sensor as illustrated by blue trace in Fig. 5.15a. Again, after approximately 60 min, saturation in optical wavelength shift was reached, this being 0.29 nm (Fig. 5.15c). The saturation in wavelength shift indicates that the CuAAC reaction has proceeded to completion for both ssDNA azide and ssDNA alkyne, while the magnitude of the shift in wavelength is comparable for both resonators, indicating a similar density of the two DNA monolayers. No binding is observed between $ssDNA_{azide}$ and the surface when using an unmodified microring, as illustrated by the green trace. This proves again site selectivity of the surface chemistry.

Multiplexed photonic sensing of DNA hybridisation

Finally, multiplexed photonic sensing was demonstrated by adding single-strand DNA complementary to $ssDNA_{azide}$ to the two photonic electrode device. For this step, we exchanged the buffer solution PB (pH7) with 1xTE 1M NaCl. The inclusion of salt screens the negative charge of the DNA in order to reduce electrostatic repulsion between DNA in the solution phase and the surface-immobilised DNA.

After injecting the complementary DNA strand to $ssDNA_{azide}$ (at 400 nM concentration), namely $cDNA_{azide}$, a corresponding shift in resonance of 0.09 nm is observed only on the micrорring functionalised with $ssDNA_{azide}$ (blue trace in Fig. 5.16a) attributable to the formation of a double-strand DNA complex (Fig. 5.16b). The lack of resonance shift on the other microring derivatised with $ssDNA_{alkyne}$ not only highlights the high spatial selectivity of the electrografting process but also confirms the multiplexed capability of the sensor. Similarly, upon exposure to 400 nM $cDNA_{alkyne}$, we observed a shift of 0.05 nm on the optical ring functionalised with $ssDNA_{alkyne}$ (Fig. 5.16a, c). No hybridisation is observed when challenging the sensor to non-complementary DNA, as illustrated by the green trace in Fig. 5.16a.

To understand the higher optical shift observed following hybridisation of $cDNA_{azide}$, it is instructive to consider the differences in the sequences of $cDNA_{azide}$ and $cDNA_{alkyne}$. $cDNA_{azide}$ contains a higher GC content than $cDNA_{alkyne}$ and is thus expected to exhibit a higher free energy of hybridisation upon binding to its complementary, surface-attached strand. From NUPACK, we calculate that the free energy of the $cDNA_{azide}$ complex is −24.2 kcal/mol compared to $cDNA_{alkyne}$ where the free energy is only −20.4 kcal/mol. The binding affinity of $cDNA_{azide}$ will thus be higher and we would expect a larger proportion of double stranded complexes compared to $cDNA_{alkyne}$. From Fig. 5.15a, we also estimate that the optical shift for probe DNA immobilised on the alkyne functionalised sensor is 15% higher than on the azide functionalised sensor. It is well known that hybridisation efficiency is dependent on the molecular density of the immobilised layer [31]. It should also

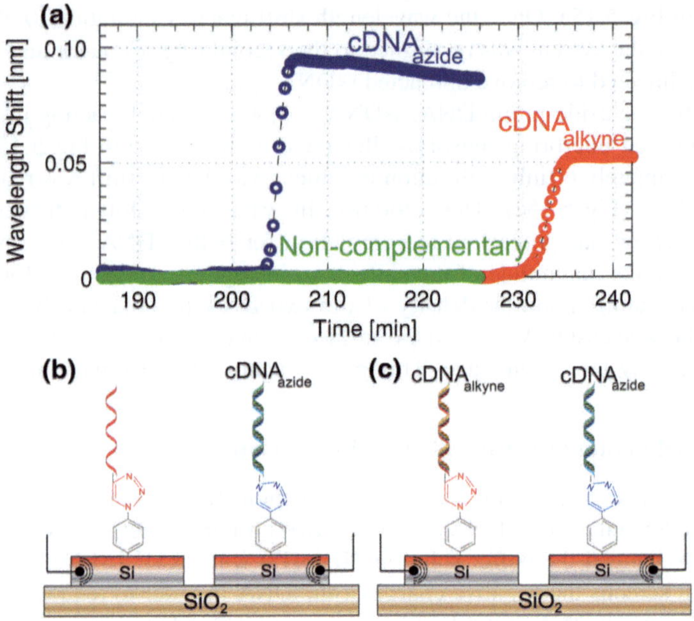

Fig. 5.16 Multiplexed photonic sensing of DNA hybridisation. **a** The selectivity of the functionalisation process is confirmed by exposing the sensor array to DNA of the reverse complement of ssDNA$_{azide}$ and ssDNA$_{alkyne}$. We note that no hybridisation is observed when challenging the sensor to non-complementary DNA, as illustrated by the green trace. **b** and **c** Injection of the reverse complements of ssDNA$_{azide}$ and ssDNA$_{alkyne}$, labelled as cDNA$_{azide}$ and cDNA$_{alkyne}$,respectively

be noted that the bioassay for the immobilisation of ssDNA$_{azide}$ and ssDNA$_{alkyne}$ (Fig. 5.15) was optimised to obtain a densely packed layer of DNA probes, and not for the detection of their complementary sequences. Therefore, the difference in the absolute shift levels between the traces represented in Figs. 5.15 ($\approx$0.29 nm) and 5.16 ($\approx$0.09 nm) can be originated from steric hindrance at the sensor surface.

5.5.3 Control Experiments for Selective DNA Hybridization

Control experiments to analyse the surface modification chemistry were performed using quartz crystal microbalance with dissipation monitoring (QCM-D) (Q-Sense E4). Gold electrodes were chosen because they were compatible with all the reactions involved. An electrochemical module (QEM 401) was fitted to the instrument in order to study and optimise the electroreduction chemistry. A constant flow rate of 20 μmL/min was kept throughout the measurements.

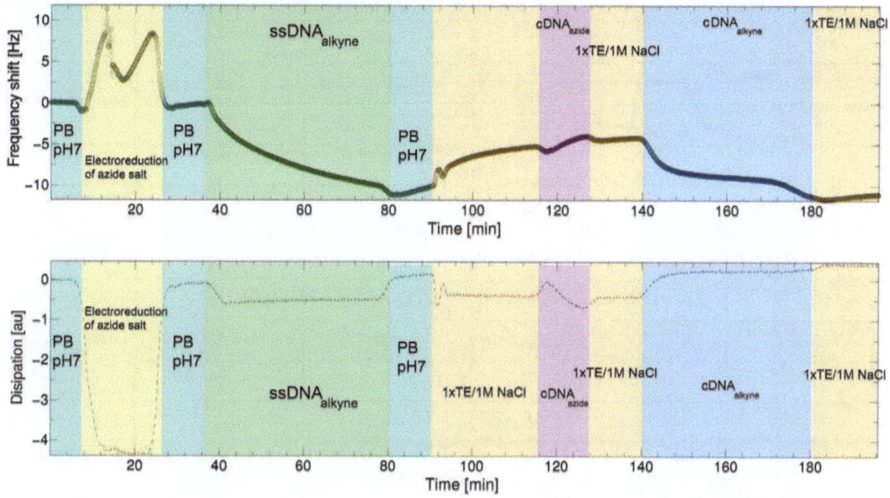

Fig. 5.17 Azide modified surface control experiment for selective DNA hybridisation. The QCM-D gold sensor was modified with ssDNA$_{alkyne}$, showing no hybridisation with cDNA$_{azide}$. In contrast, hybridisation is observed when flowing the complementary cDNA$_{alkyne}$

Firstly, the QCM-D gold sensor was modified using diazotisation and electrografting of 4-azidoaniline (Fig. 5.17). ssDNA$_{alkyne}$ (1 μM in PB (pH 7)) was then injected and immobilised to the surface through azide-alkyne Huisgen cycloaddition (same mixture as in Sect. 5.5.2), indicated by the resonance shift of 10 Hz 45 min from injection. Thereafter, the sensor was rinsed with PB (pH 7) and the buffer was exchanged to 1 × TE 1 M NaCl to decrease electrostatic repulsion as discussed previously in Sect. 5.5.2. cDNA$_{azide}$ was injected (400 nM), and no change in mass or dissipation was observed following injection. This indicated no hybridisation with the strand fixed to the surface (ssDNA$_{alkyne}$) and cDNA$_{azide}$. In contrast, hybridisation is observed when flowing the complementary cDNA$_{alkyne}$ (400 nM) through the sensor, hence demonstrating selective DNA sensing.

Additionally, a control experiment was performed to examine if deterioration of the chemical activity of the surface alkyne after exposure to the chemicals required for electroreduction of azidoaniline was produced (Fig. 5.18). In this case, the QCM-D gold sensor was modified by diazotization and electrografting of 4-ethynylbenzene diazonium. The modified surface was next exposed to the solution required for azide modification, as required during bi-functionalisation of the photonic array. ssDNA$_{azide}$ (1 μM in PB (pH 7)) was then injected and immobilised to the surface through azide-alkyne Huisgen cycloaddition (same mixture as in Sect. 5.5.2). The increase in mass is consistent with the assembly of a ssDNA$_{azide}$ layer and suggests no deterioration of the chemical activity of the alkyne surface after exposure to the chemicals required for electroreduction of 4-azidoaniline. At this stage the

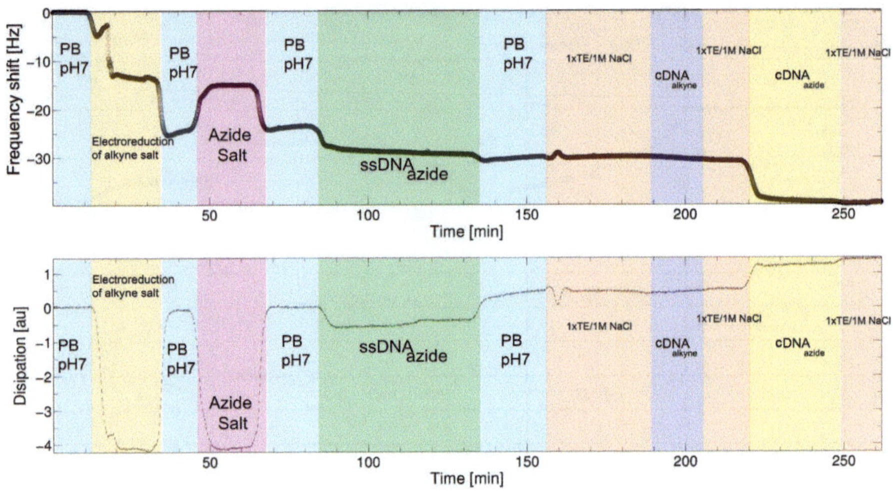

Fig. 5.18 Alkyne modified surface control experiment for selective DNA hybridisation. The QCM-D gold sensor was modified with ssDNA$_{azide}$ after the surface had been exposed to the solution required for azide modification. No hybridisation with cDNA$_{alkyne}$ was observed, while hybridisation was observed when flowing the complementary cDNA$_{azide}$, suggesting no deterioration of the chemical activity of the alkyne surface after exposure to the chemicals required for electroreduction of 4-azidoaniline

sensor was rinsed with PB (pH 7) and the buffer was exchanged to $1 \times$ TE 1 M NaCl. No change in mass or dissipation was observed following injection of non-complementary cDNA$_{alkyne}$ (400 nM concentration in $1 \times$ TE 1M NaCl), indicating no hybridisation with the strand fixed to the surface. In contrast, hybridisation is observed when flowing the complementary cDNA$_{azide}$ (400 nM in $1 \times$ TE 1M NaCl) through the sensor, hence demonstrating selective DNA sensing.

5.6 Summary

We have demonstrated how our technology can not only be applied to the study of electrochemical reactions, but is also capable of multiplexed detection at the microscale, and possibly even on sub-micrometer scales. The electrografting of diazonium ions of 4-ethynylbenzene diazonium, n-(4-aminophenyl)maleimide and 4-azidoaniline has been studied, which provide either alkyne, maleimide or azide modification of the surface, respectively. We have been able to provide insight into the structure of these layers, calculating the surface density of molecules for each of them and discerning between highly packed monolayers and multilayer assemblies. Examples of the high versatility of functionalisation with electrografted layers of diazonium ions have also been demonstrated, here being the immobilisation of the electrochemical probes methylene blue and ferrocene through thiol-maleimide reaction and Copper-Catalyzed Azide-Alkyne Huisgen Cycloaddition click reaction, respectively. Finally, we demonstrated how electrografting of diazonium ions can pave a route for the development of high density arrays of photonic biosensors. We have created a two electro-photonic ring resonator device capable of multiplexed detection through site-selective modification of the surface chemistry, enabling site specific detection of DNA-bioconjugation.

References

1. J. Pinson, F. Podvorica, Attachment of organic layers to conductive or semiconductive surfaces by reduction of diazonium salts. Chem. Soc. Rev. **34**, 429–439 (2005)
2. J. Pinson, M. Delamar, R. Hitmi, J. Michel, Covalent modification of carbon surfaces by grafting of functionalized aryl radicals produced from electrochemical reduction of diazonium salts. J. Am. Chem. Soc. **114**(14), 5883–5884 (1992)
3. S. Gam-Derouich, C. Mangeney, S. Mahouche-Chergui, M. M. Chehimi, Aryl diazonium salts: a new class of coupling agents for bonding polymers, biomacromolecules and nanoparticles to surfaces. Chem. Soc. Rev. **40**, 4143–4166 (2011)
4. T. Waitz, T. Prohaska, W. Lindner, H. Hinterwirth, S. Kappel, M. Lmmerhofer, Quantifying thiol ligand density of self-assembled monolayers on gold nanoparticles by inductively coupled plasmamass spectrometry. ACS Nano **7**(2), 1129–1136 (2013)
5. L.F. Policar, D. Evrard, V. Balland, B. Limoges, Electrochemical functionalization of carbon surfaces by aromatic azide or alkyne molecules: A versatile platform for click chemistry. Chem. A Eur. J. **14**(30), 9286–9291 (2008)
6. S.H. DuVall, F. Anariba, R. L. McCreery, Mono- and multilayer formation by diazonium reduction on carbon surfaces monitored with atomic force microscopy scratching. Anal. Chem. **75**, 3837–3844 (2003)
7. D.J. Garrett, V. Nock, B.S. Flavel, A.J. Gross, A.J. Downard, A simple approach to patterned protein immobilization on silicon via electrografting from diazonium salt solutions. ACS Appl. Mater. Interfaces **2**(4), 1184–1190 (2010)
8. Robert R. Gagne, Carl A. Koval, George C. Lisensky, Ferrocene as an internal standard for electrochemical measurements. Inorg. Chem. **19**(9), 2854–2855 (1980)
9. A. Michael, On the addition of sodium acetoacetate- and sodium malonic acid esters to the esters of unsaturated acids. Am. Chem. J. **9**, 115 (1887)
10. T. Miyadera, E.M. Kosower, Receptor site labeling through functional groups. 2. reactivity of maleimide groups. J. Med. Chem. **15**(5), 534–537 (1972)
11. D. Langford, R. Bonand, E. Koutsoumpeli, J. Murray, S. Johnson, Probing molecular interactions with methylene blue derivatized self-assembled monolayers. Sens. BioSensing Res., 1–6 (2015)
12. H. Kolb, M.G. Finn, K.B. Sharpless, Click chemistry: Diverse chemical function from a few good reactions. Angew. Chem. Int. Ed. **40**, 2004–2021 (2001)
13. I.S. Choi, Y.S. Chi, Surface reactions on demand: Electrochemical control of sam-based reactions. Angew. Chem. Int. Ed. **45**(30), 4894–4897 (2006)
14. Y. Zhao, X. Zhao, G. Zhongze, Photonic crystals in bioassays. Adv. Funct. Mater. **20**(18), 2970–2988 (2010)
15. W. C. Lai, C.Y. Lin, Y. Zou, S. Chakravarty, R.T. Chen, Methods to array photonic crystal microcavities for high throughput high sensitivity biosensing on a silicon-chip based platform. Lab Chip, **12**, 2309–2312 (2012)

16. Genalyte (2009), http://www.http://genalyte.com/
17. B. Spaugh, F. Tybor, W.G. Gunn, M. Hochberg, T. Baehr-Jones, R.C. Bailey, M. Iqbal, M.A. Gleeson, L.C. Gunn, Label-free biosensor arrays based on silicon ring resonators and high-speed optical scanning instrumentation. IEEE J. Sel. Top. Quantum Electron. **16**(3), 654–661 (2010)
18. A. Kazmierczak, F. Dortu, M.J. Banuls Polo, A. Maquieira Catala, G.M. Kresbach, T. Sohlstreom, T. Moh, L. Vivien, J. Popplewell, G. Ronan, A. Barrios, G. Stemmea, K.B. Gylfason, W. van der Wijngaarta, A packaged optical slot-waveguide ring resonator sensor array for multiplex label-free assays in labs-on-chips. Lab Chip **10**(5), 281–290 (2010)
19. V. Sanz, J. Berthelot, J. L. Garcia-Cordero, J. Renger, S. J. Maerkl, M. P Kreuzer, S. Acimovic, M. A. Ortega, R. Quidant, Lspr chip for parallel, rapid, and sensitive detection of cancer markers in serum. Nano Lett. **14**(5), 2636–2641 (2014)
20. M. Soler, M. C. Estevez, M. Alvarez, M. A. Otte, B. Sepulveda, L. M. Lechuga, Direct detection of protein biomarkers in human fluids using site-specific antibody immobilization strategies. Sensors, **14**(2), 2239 (2014)
21. J.W. Chamberlain, E.D. Christensen, M. HochbergD, M.J.T. Kirk, G.E. Fridley, D.M. Ratner, Multiplexed inkjet functionalization of silicon photonic biosensors. Lab Chip **11**(7), 1372–1377 (2011)
22. A. J. Qavi, M. S Luchansky, A. L. Washburn, M. S. McClellan, J. T. Kindt, R. C. Bailey. Silicon photonic microring resonator arrays for scalable and multiplexable bioanalysis. Proc. SPIE, **7888**, 788802–788802–4 (2011)
23. V.M. Kurt, M.A. Ratner, Electron tunneling in solid-state electron-transfer reactions. Chem. Rev. **87**(1), 113–153 (1987)
24. S. Palacin, J. Leroy, F. D. Del, L. Zagonel, N. Barrett, O. Renault, A. Bailly, J. Charlier, D. Mariolle, Local silicon doping as a promoter of patterned electrografting of diazonium for directed surface functionalization. J. Mater. Chem. **18**, 3136–3142 (2008)
25. J. Leroy, D. D. Frari, Z. Doriane, N. Barette, O. Renault, A. Bailly D. Mariolle J. Charlier, S. Palacin, Local silicon doping as a promoter of patterned electrografting of diazonium for directed surface functionalization. J. Mater. Chem. **18**, 3136–3142 (2008)
26. C. Bureau, J. Charlier, E. Clolus, S. J. Palacin, Selectivity of organic grafting as a function of the nature of semiconducting substrates. J. Electroanal. Chem. **625**(1), 97–100 (2009)
27. M.J. Heller, DNA microarray technology: Devices, systems, and applications. Annu. Rev. Biomed. Eng. **4**, 129–153 (2002)
28. H-Y. Wu, F. Yang, N. Ganesh, D. O. Gonzalez, G. Bollero, L. O. Vodkin P. C. Mathias, S. I. Jones, B. T. Cunningham, Improved sensitivity of dna microarrays using photonic crystal enhanced fluorescence. Anal. Chem. **82**(16), 6854–6861 (2010)
29. J.S. Bois, B.R. Wolfe, M.B. Pierce, A.R. Khan, R.M. Dirks, J.N. Zdeh, C.D. Steenberg, N.A. Pierce, Nupack: Analysis and design of nucleic acid systems. J. Comput. Chem. **32**(1), 170–173 (2011)
30. I. S. Presolski, Vu Phong Hong, M.G. Finn, *Copper-Catalyzed AzideAlkyne Click Chemistry for Bioconjugation*. Wiley Inc. (2009)
31. A.J. Heeger, K.W. Plaxco, F. Ricci, R.Y. Lai, J.J. Sumner, Effect of molecular crowding on the response of an electrochemical DNA sensor. Langmuir **23**(12), 6827–6834 (2007)

Chapter 6
Tailoring Light-Matter Interaction for Quantification of Biological and Molecular Layers

The silicon electro-photonic platform was introduced in Chap. 4, while its capabilities for monitoring electrochemical reactions in situ and site-selective functionalisation were presented in Chap. 5. By tailoring the light-matter interaction, it is also possible to create a sensor with the potential to enable detailed quantification of biological and molecular layers. Here I present such a novel photonic sensor that consists of multiple ring resonators with different distributions of the optical field that enables a measurement of both thickness and refractive index of an immobilised layer. Such a technology uniquely enables measurement of conformational changes of bound molecules. As proof-of-concept, the sensor is applied to the study of exemplar immobilised layers and the conformational change of a hairpin-shape DNA strand.

6.1 Introduction

In optical biosensors, the interaction between light (optical field) and matter (biological molecules) leads to changes in at least one property of the sensor. For instance, in optical ring resonator sensors, the overlap of the optical field with surface-bound molecules modifies the resonance wavelength of the cavity [1]. Similarly, in sensors based on surface plasmon resonance (SPR) [2], adsorbed molecules change the resonance conditions of the surface plasmon waves.

In SPR, the light-matter interaction is fixed, being a direct result of the nature of the sensing principle. In contrast, sensors based on photonic ring resonators can be engineered to control the light-matter interaction. For example, by controlling the geometry of the sensor or the light polarisation, the distribution of the optical field can be tailored to provide a particular optical profile and penetration depth. While the focus here is on ring resonators, such control can be achieved in a wide range of photonic transducers such as PhCs [3] or Mach-Zehnder interferometers [4].

J. Juan Colaś , *Dual-Mode Electro-photonic Silicon Biosensors*, Springer Theses,
DOI 10.1007/978-3-319-60501-2_6

Here, I present a sensor technology which consists of two cascaded ring resonators whereby the geometry of each ring is controlled and optimised so the tail of each optical field in the solution-phase has a different distribution. One ring is engineered to exhibit a short penetration depth and thus increased sensitivity to changes in the refractive index local to the interface between the waveguide and the solution. While the second resonator is designed with a large tail to penetrate further into the solution. By combining the optical response of the two resonators, it is possible to provide quantitative information about the thickness and refractive index (and hence density) of a surface layer. In the following section, I explain how this combination of sensors can be exploited to provide insight into the structure of immobilised molecules.

6.2 Geometry Control on Silicon Electro-Photonic Biosensors: The Cascaded Ring Resonator Configuration

Most evanescent field based sensors, including SPR, only measure the interaction of a single energy distribution of an optical mode with a biological or molecular layer. Such a sensor can only provide information about bound mass or the concentration in a solution. As a result, they only provide information on the combined change of the refractive index and thickness of the layer, without being able to distinguish between these two effects. The sensor is thus largely insensitive to changes in shape or conformation of bound biomolecules. A disentanglement of these parameters would yield critical information on the conformation of biomolecules [5], or on the structure of a biological or molecular layer. This additional information could be obtained using a sensor that can accurately measure both the thickness (t) and the refractive index (n) of a biolayer, simultaneously.

An example of a commercial label-free biosensor that can accurately distinguish between n and t layer is the dual polarisation interferometer (DPI) (Farfield Scientific Ltd.) [6] (the sales of this instrument were discontinued in 2012). DPI is an interferometer-based sensor in which two orthogonally polarised modes interact with the molecules. Molecular binding will impact differently on the effective refractive index of each of the modes due to their different profiles. This provides two equations for the two unknowns, t and n. A unique solution for t and n can thus be calculated from the intersection of these two equations. Such a dual-polaristion approach has also been implemented in a silicon ring resonator sensor [7], showing similar performance to DPI.

These two examples require the transducer to be excited with two optical modes simultaneously. Such an approach is however not trivial or even possible in some structures. For example, in [7], an asymmetrical directional coupler is required to excite both TE and TM polarisations simultaneously in the ring cavity. Here, we present a simpler approach to this challenge, which employs a single TE mode to excite two independent ring resonators with a different waveguide width. The

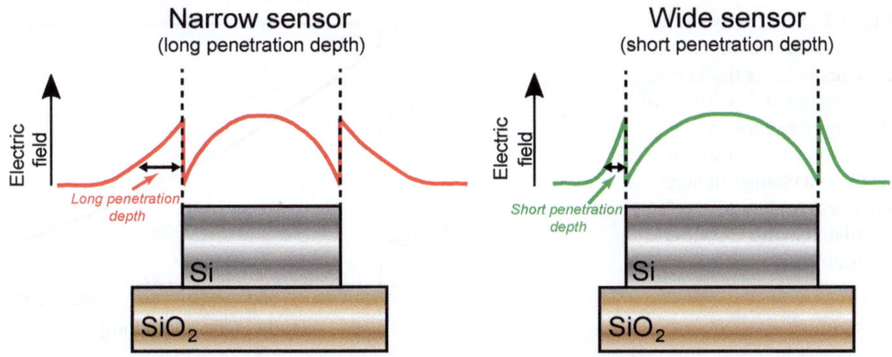

Fig. 6.1 Cross section profile of the proposed dual ring resonator scheme. By modifying the profile of the optical field in the solution-phase, each ring will respond differently to changes at the interface, hence providing the required two-equation system for monitoring t and n of overlayers

different geometry of the two resonators changes the distribution of the optical field in the solution-phase. The resonance wavelengths of each ring will thus shift differently when biomolecules interact at the sensor surface to provide the required two-equation system. A sketch of the cross section of the waveguides of these two rings is shown in Fig. 6.1.

6.2.1 *Cascaded Ring Design and Fabrication*

As discussed in Sect. 2.3.2.4, the resonance wavelength of the TE mode in a microring cavity is determined by Eq. 2.11. The mode senses the biolayer at the surface by means of the evanescent tail that penetrates into the solution-side of the interface waveguide/solution. Changes at the surface of the ring resonator result in a change in n_{eff}, detected by a shift in the resonance wavelength, as:

$$\Delta\lambda_{resonance}(t,n) = \frac{\lambda_{resonance} \cdot \Delta n_{eff}(t,n)}{n_g} \tag{6.1}$$

Here $\Delta n_{eff}(n,t)$ is the change in effective refractive index of the optical mode due to changes in the cavity environment, while $\lambda_{resonance}$ and n_g account for the resonance wavelength and the group index, respectively. Figure 6.2 shows the simulated sensitivity of the ring resonator as a function of the waveguide width to a change in layer thickness ($\frac{\partial\lambda_{resonance}}{\partial t}$), with t the thickness of the biolayer. The sensitivity is expressed in nm/nm, and the simulation was performed using COMSOL Multiphysics using a molecular layer with a refractive index of 1.45 RIU [8]. The thickness of the waveguide was fixed at 220 nm and the resonance wavelength was 1550 nm. It shows how waveguides that have different width have different sensitivity to changes in the thickness of the layer.

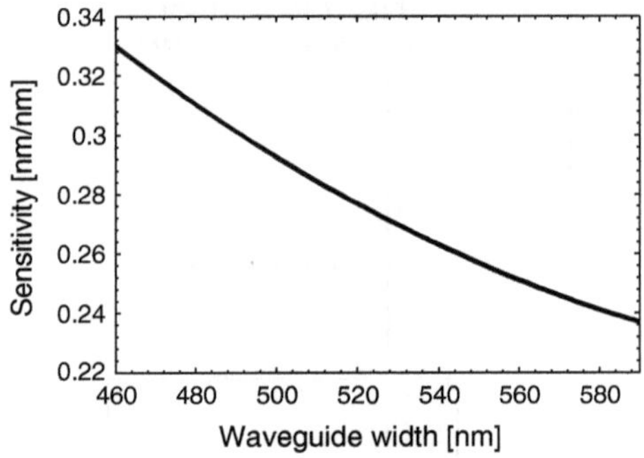

Fig. 6.2 Sensitivity of the ring resonator with respect to changes in layer thickness. The simulated sensitivity of the ring resonator as a function of the waveguide width to a change in layer thickness is shown, simulated using COMSOL Multiphysics

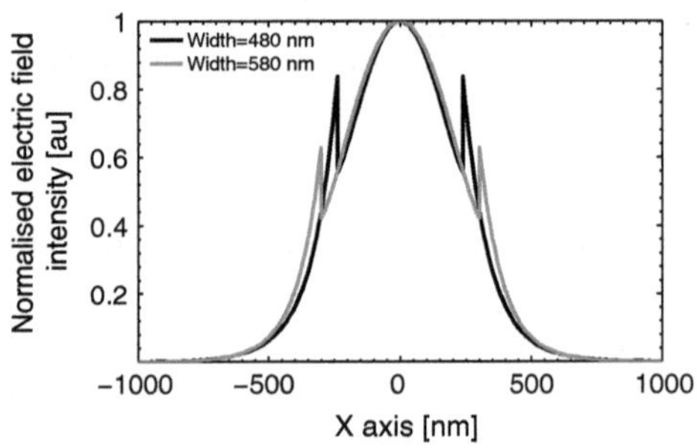

Fig. 6.3 Energy distribution of the optical mode for waveguides of 480 nm and 580 nm. The evanescence tail extends 80 nm into the cladding for the 480 nm width waveguide, while it extends 49 nm for the 600 nm width waveguide

The ideal cascaded ring sensor consists of sensors in which the optical mode profile has maximum difference to increase the orthogonality of the two-equation system. Such a large difference can be achieved using resonators with a large difference in width. However, there are two main limitations to consider. Firstly, guidance of second order modes needs to be avoided as they decrease the Q-factor of the cavity. This condition is satisfied for a width of the cavity waveguide smaller than 586 nm [9]. For ring widths bigger than 360 nm, the propagation loss decreases, again leading to a reduction in Q-factor of the cavity [10]. These considerations lead to a choice of waveguide width between 475–580 nm. Figure 6.3 shows a simulation (using COMSOL Multiphysics) of the energy distribution of the two waveguides of 480 and 580 nm width (considering water as cladding and at the resonance wavelength of 1550 nm).

The penetration depth is defined as the distance over which the energy density of the modes decreases by a factor 1/e in the solution-phase, which amounts to approximately 80 and 49 nm for 480 and 580 nm waveguide widths, respectively. Therefore, after validation from simulations, we chose the waveguide widths of 480 and 580 nm for our sensors.

We designed the sensors with a cavity length of approximately 200 μm and 210 μm, respectively for the narrow (480 nm) and the wide (580 nm) sensor. Different cavity lengths were employed to avoid overlap between coinciding resonances. The device was fabricated with the same rib waveguide profile and following the procedure detailed in 4.4. The final fabricated device is shown in Fig. 6.4.

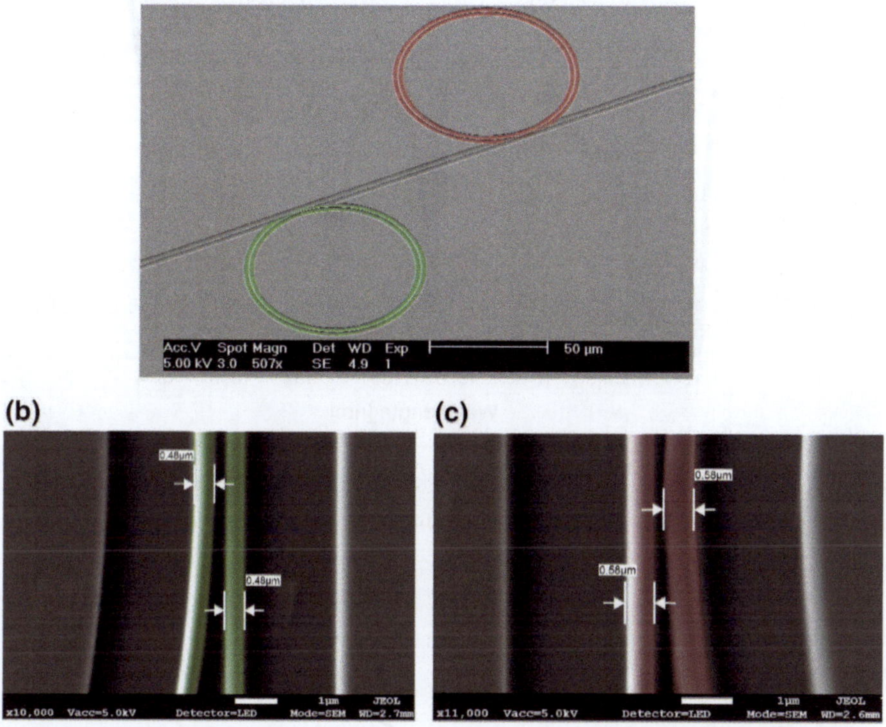

Fig. 6.4 Cascaded ring configuration. **a** SEM micrograph of two ring resonators placed next to each other exhibiting a different optical mode profile due to the waveguide width difference. **b** Close-up SEM image of the narrow sensor (480 nm). **c** Close-up SEM image of the wide sensor (580 nm)

The wavelength range of the optical spectrum between 1546 nm and 1555 nm measured at the output waveguide is shown in Fig. 6.5, which clearly displays the resonance dips of both cavities.

Due to the difference in cavity length, and in agreement with Eq. 2.14, the two rings exhibit different FSR and resonance wavelength. The measured $FSRs$ are 2.42 nm and 2.69 nm for the narrow and wide sensors, respectively. The coupling section between the access waveguide and the ring cavity was also optimised for high Q-factors in both cavities, leading to gap lengths of 230 nm and 180 nm, respectively for the narrow and wide sensor. The obtained Q-factors with air cladding of the cavities were approximately 50,000 and 40,000 (480 nm and 580 nm waveguide width resonators, respectively).

As in Sect. 4.4, the bulk sensitivity of each device was calculated empirically by tracking the resonance wavelength of the device when challenged with a range of NaCl solutions of different molar concentrations. Again, the refractive index of each solution was calculated as detailed in Sect. 4.4. The result of this experiment is shown in Fig. 6.6, displaying a clear linear relationship between resonance shift and

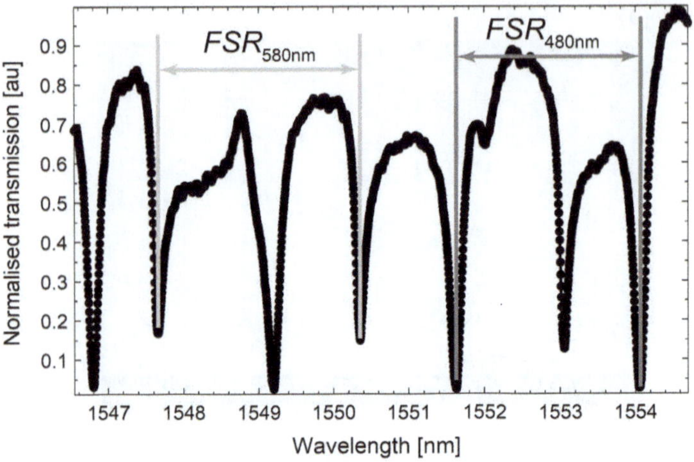

Fig. 6.5 Transmission spectrum of the cascaded ring configuration (air cladding). Q-factors of 50,000 and 40,000 were obtained for the cavities, exhibiting FSR of 2.42 nm and 2.69 nm, respectively for the 480 and 580 nm waveguide width sensors

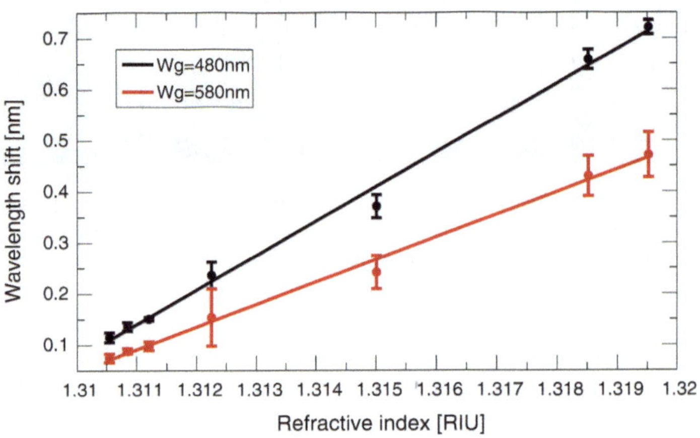

Fig. 6.6 Resonance wavelength shift against changes in the bulk refractive index of the cascaded ring configuration. Sensitivities of 65 nm/RIU and 43 nm/RIU are obtained (equivalent to *LODs* of 1.15×10^{-5} RIU) and 2.45×10^{-5} RIU

solution concentration for each sensor. The gradients of the lines were 65 nm/RIU and 43 nm/RIU. The minimum resolvable optical shift for each ring was calculated to be 0.75 pm and 1.05 pm for the 480 and 580 nm waveguide, respectively (as described in 4.4). These minimum resolvable optical shifts, together with the sensitivity of the sensor, lead to *LODs* of 1.15×10^{-5} RIU and 2.45×10^{-5} RIU (in good agreement with similar ring resonators [1, 10, 11]).

6.3 Extraction of Thickness and Refractive Index

A two-equation system linking the wavelength shift for each sensor ($\Delta\lambda_{480nm}$ and $\Delta\lambda_{580nm}$, for the narrow and wide ring, respectively) is required to extract t and n of the immobilised layer, a:

$$\Delta\lambda_{480nm} = f(t, n) \tag{6.2}$$

$$\Delta\lambda_{580nm} = g(t, n) \tag{6.3}$$

Here, $f(t, n)$ and $g(t, n)$ are nonlinear functions relating the interaction of the optical field with the overlayer. In order to determine the functions $f(t, n)$ and $g(t, n)$, a set of electromagnetic simulations using COMSOL Multiphysics where performed to calculate $\Delta\lambda_{480nm}$ and $\Delta\lambda_{580nm}$ as a function of t and n. These were subsequently fitted to the analytical model reported in [7], which describes the optical wavelength shift as a function of t and n of an overlayer on the resonator waveguide. This pseudo 2D analytical model was obtained by using standard waveguide theory to solve the perturbation of the refractive index profile due to the overlayer. t and n relate to the thickness and refractive index of the overlayer, and n_b is the refractive index of the buffer solution. The wavelength shift $\Delta\lambda(t, n)$ can then be described as follows:

$$\Delta\lambda(t, n) = B\frac{(n - n_b)nf_p(t, n)}{1 + n^2 f_p(t, n) + n_b^2 f_b(t, n))} \tag{6.4}$$

$$f_p(t, n) = A_{t,p}[1 - exp(-2\delta_p t)] + A_{s,p}[1 - exp(-2\delta_p t)] \tag{6.5}$$

$$f_b(t, n) = A_{t,b}[exp(-2\delta_b t)] + A_{s,b}[exp(-2\delta_b t)] \tag{6.6}$$

$$\delta_p = k(n_{eff} - n)^{-1/2} \tag{6.7}$$

$$\delta_b = k(n_{eff} - n_b)^{-1/2} \tag{6.8}$$

The functions f_p and f_b are the contributions from the overlayer and the buffer, respectively. Each of these has a contribution from the top surface of the waveguide, indexed with t, and a contribution from the sidewalls, indexed with s. The normalised amplitudes at the interface between buffer and overlayer $A_{t,b}$ and $A_{s,b}$ are determined by $A_{t,p}$ and $A_{s,p}$ using boundary conditions, such that there are four parameters to fit per waveguide geometry: n_{eff}, $A_{t,p}$, $A_{s,p}$ and B. To obtain these parameters, the wavelength shift of both wave guide geometries is simulated for $t = 0 - 20$ nm and $\Delta n = 0 - 0.2$ RIU (difference between n and n_b) for a rib etched waveguide as

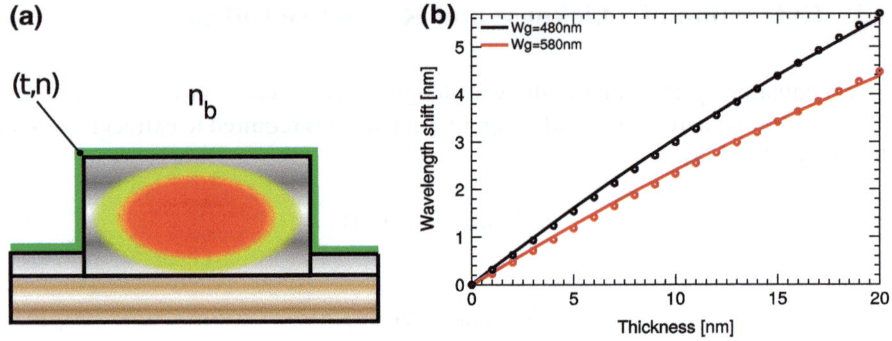

Fig. 6.7 Modelling of wavelength shifts originated from overlayers on the ring sensors. **a** Cross section profile of the ring waveguide as used for simulations. **b** Simulations of the wavelength shifts for both waveguide geometries in function of the thickness of the overlayer (refractive indexes of 1.45 RIU for the overlayer and 1.32 RIU for the buffer)

depicted on Fig. 6.7a. These simulated shifts can then be fitted to the analytical model using a least-square metric with physical constraints on the parameters (Fig. 6.7b). A coefficient of determination R^2 of 0.9992 is obtained, confirming the validity of this model for our structure. Here I assumed a layer of homogeneous refractive index and constant thickness.

Once the four parameters of the model have been retrieved for each waveguide width, it is possible to solve the measured wavelength shifts $\Delta\lambda_{480nm}$ and $\Delta\lambda_{580nm}$ through the obtained functions $f(t, n)$ and $g(t, n)$ to t and n. By setting $f(t, n)$ equal to $\Delta\lambda_{480nm}$ and $g(t, n)$ to $\Delta\lambda_{580nm}$, two implicit functions in t and n are obtained. Figure 6.8 shows a plot of these functions for a layer with thickness of 3.5 nm and refractive index of 1.485 RIU. The intersection point of both curves corresponds to the solution of the experiment uniquely solved for the overlayer properties, given the measured $\Delta\lambda_{480nm}$ and $\Delta\lambda_{580nm}$ shifts.

The noise of this configuration have been quantified as three times the standard deviation on a linear curve through calibration measurements (similarly as performed in 4.4). Experimentally, the noise equals 0.43 nm for the 480 nm waveguide width sensor and 0.38 nm for the 580 nm waveguide. I note that these noise levels are in good agreement with the noise determined in [7], where a similar ring sensor is used. The maximum error in determining t and n is calculated based on Gaussian wavelength noise with 0.43 nm standard deviation for both sensors. This results in an error of 0.05 nm for the thickness estimation and in 0.005 RIU for the refractive index calculation. These errors are determined for an overlayer with $n = 1.45$ RIU and $t = 2$ nm.

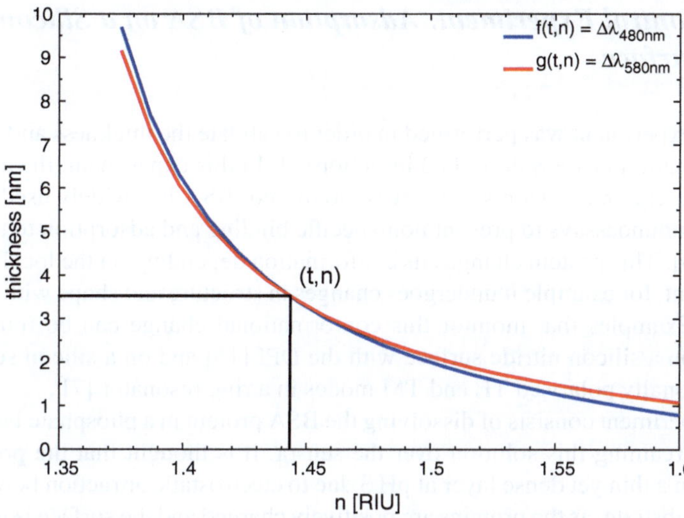

Fig. 6.8 Solving protocol by plotting $f(t,n) = \Delta\lambda_{480nm}$ and $g(t,n) = \Delta\lambda_{580nm}$. The intersection point of both curves corresponds to the solution of the experiment

6.4 Cascaded Rings for Characterising Biological and Molecular Monolayers

In this section we report three experimental proof-of-concept experiments using the cascading ring configuration to measure the refractive index and thickness of three different molecular layers. Firstly, the adsorption of Bovine Serum Albumin (BSA) protein on the silicon surface was monitored. Secondly, a layer of MPTS (Sect. 2.5.2), which has been extensively employed in this thesis to provide a thiol modification to the silicon surface. Thirdly, a layer of bioreceptors, known as Affimers (sub-variant of peptide aptamers [12]) selective against the protein IgG2a.

To accurately extract t and n from $\Delta\lambda_{480nm}$ and $\Delta\lambda_{580nm}$ via the simulations performed in Sect. 6.3, the exact geometry of the waveguide needs to be known. The dimensions of the device was thus characterised by AFM (BioScope Resolve 2, Bruker). Geometries of 482 × 202 nm and 578 × 205 nm were respectively obtained for the width and height of each resonator. The refractive index of the buffer (n_b) employed in the first experiment was obtained from [7], it being 1.3163 RIU; while for the second and third experiments, being 1.3505 RIU for ethanol and 1.319 RIU for Phosphate Buffer Saline (PBS, pH 7.4), were obtained from [8].

6.4.1 Control Experiment: Adsorption of BSA on a Silicon Surface

A control experiment was performed in order to validate the thickness and refractive index estimation process described insection 6.3. In this experiment, the adsorption of BSA protein on a silicon surface was monitored. BSA is a widely used blocking agent in immunoassays to prevent non-specific binding and adsorption to the sensor surface [13]. This protein changes its conformation depending on the local chemical environment, for example it undergoes changes in structure and shape with changing pH [14]. Examples that monitor this conformational change can be found in the literature on a silicon nitride surface with the DPI [15] and on a silicon surface via the orthogonally polarised TE and TM modes in a ring resonator [7].

The experiment consists of dissolving the BSA protein in a phosphate buffer solution and streaming this solution over the sensor. It is thought that the protein will assemble in a thin yet dense layer at pH 3 due to electrostatic attraction between proteins and substrate, as the proteins are positively charged and the surface is negatively charged [7].

The pH 3 phosphate buffer solution was prepared at a concentration of 2 mM, to which BSA proteins were added until a concentration of 0.1 mg/mL was obtained. Firstly, phosphate buffer pH 3 was used and flowed over the sensor (at a rate of 30 μL/min) until a stable resonance wavelength was obtained for each sensor (Fig. 6.9a). Subsequently, BSA in phosphate buffer pH 3 was flowed over the sensor (at 30 μL/min) for 60 min and the left incubating for 3 min.

Figure 6.9 shows the results obtained for the adsorption of BSA at pH 3. In Fig. 6.9a, at minute 24, the resonance wavelength of each sensor begins to shift as BSA proteins are adsorbed on the silicon surface. The estimation of both thickness and refractive index is thereafter performed as shown in Figs. 6.9, **b** and **c**, respectively. In this case, we obtained a layer thickness of 1.26 nm +/− 0.05 nm and a refractive index of 1.43 +/− 0.005 RIU, which are both in good agrement with the values estimated with a dual polarisation ring resonator (thickness of 1.4 nm and refractive index of 1.433 RIU) [7]. The fact that we obtained similar values to those reported in [7] using the same silicon surface confirms our thickness and refractive index estimation process.

However, I note the low yield of this experiment as discussed in [16]. Furthermore, it should be noted that phosphate buffers provide effective buffering in the pH range of 6.4 to 7.4 [17], thus its buffering capacity is highly diminished at the pHs used in this experiment. Therefore the bioassay for this experiment should be optimised to obtain higher yield, which is beyond the scope and goal of this thesis.

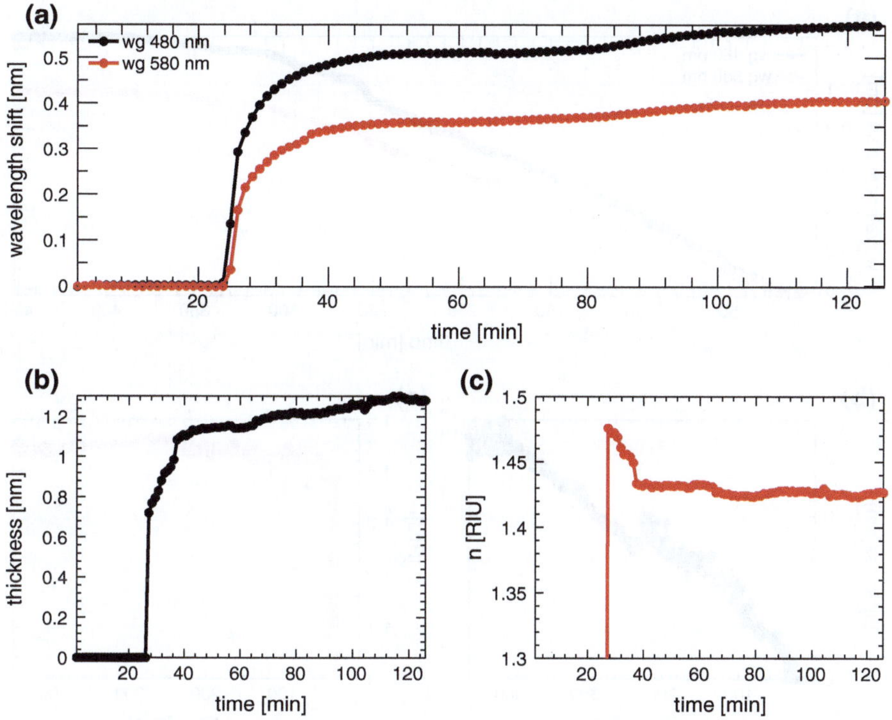

Fig. 6.9 Monitoring of the adsorption of BSA on a silicon surface. **a** The resonance wavelength of both 480 and 580 nm width ring sensors was continuously monitored while BSA molecules are adsorbed on the sensor surface at pH 3. **b** and **c** show the estimations of both thickness and refractive index of the layer, respectively

6.4.2 Characterisation of a MPTS SAM

As discussed in Sect. 2.5, SAMs are a critical component of surface based biosensors due to their ability to modify and control the chemistry and properties of planar surfaces, as they form a well-ordered and stable thin film. In this section we show how our cascaded ring configuration can provide insight into the layer assembly of the MPTS SAM discussed in Sect. 2.5.2, demonstrating the application of this technique for fundamental studies such as SAM formation.

The formation of a MPTS layer on the silicon waveguide of our device was monitored by continuous tracking of the resonance wavelength of both the 480 and 580 nm width ring sensors (Fig. 6.10a). 10 mL of ethanol containing 400 μL of MPTS (Sect. 2.5.2) was flowed through the PDMS channel. Once the solution completely filled the microfluidic channel, the flow was stopped and the reaction was incubated while the resonances of both cavities were tracked continuously.

As shown in 6.10a, formation of the layer was observed until approximately 400 min from the injection time ($\Delta\lambda_{480nm}$ = 1.28 nm and $\Delta\lambda_{580nm}$ = 1.05 nm),

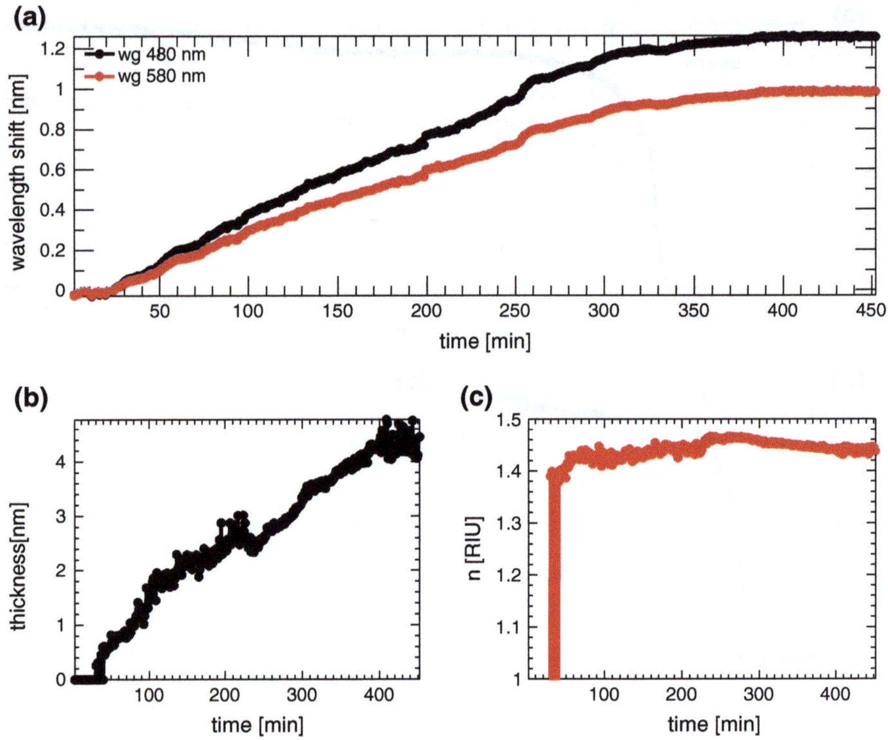

Fig. 6.10 Monitoring of a MPTS layer formation. **a** The resonance wavelength of both 480 and 580 nm width ring sensors was continuously monitored. **b** and **c** show the estimations of both thickness and refractive index of the layer, respectively

after which the shift in resonance frequency saturates. This matches with the slow reorganisation process (on the order of hours) required for SAMs to maximise the density of molecules and minimise the defects in the SAM [18]. The thickness of the assembled monolayer is found to be of 4.32 +/− 0.05 nm as shown in Fig. 6.10b. This is higher than those obtained with monolayer of SAMs (typically between 1–3 nm [18]), and indicates that a multilayer structure is being assembled on the surface rather than a well-ordered monolayer. The information obtained from the refractive index calculation (Fig. 6.10c) also supports this argument, as here a value of 1.431 +/− 0.005 RIU is obtained which is far from those values typically reported for SAMs ($n_{SAM} \approx 1.55$ RIU [18]). This difference in the refractive index value is indicator of a non-homogeneous and low-density layer.

I note that the noise displayed throughout all the traces in Fig. 6.10 is due to the low output signal power obtained from the sample used in this experiment due to a sub-optimal angled facet leading to high coupling losses.

6.4.3 Characterisation of an Affimers Biolayer

Affimers (Avacta Life Sciences Ltd.) are robust and stable antibody mimetics that are ideally suited as selective bioreceptors in biosensors [19]. Here we quantify a layer of these bioreceptors sensitive to the protein IgG2a. This particular Affimer is engineered with a cysteine at the C-terminus so it can be covalently bonded to a thiol reactive surface, for example maleimide or gold surfaces. Here, we employed the silane chemistry detailed in 4.6 to render the surface thiol reactive and to immobilise the IgG2a Affimer.

The silicon surface was firstly thiol modified by forming a MPTS layer (Sect. 2.5.2), as performed in the previous Sect. 6.4.2. Secondly, a solution of Cu^{2+} ions was injected to be able to further functionalise the surface with cysteine modified molecules, as explained in Sect. 2.5.2. The surface was then washed with PBS (pH 7.4) until unreacted Cu^{2+} ions were removed and a stable resonance wavelength was obtained for both sensors. Finally, a solution of IgG2a Affimer reagents at a concentration of 10 μg/mL was injected at 20 μL/min while both resonance wavelengths were continuously monitored (Fig. 6.11, A)).

As shown in Fig. 6.11a, the Affimer immobilisation process started after approximately 20 min from the beginning of the experiment. Saturation (with $\Delta\lambda_{480nm}$ = 1.12 nm and $\Delta\lambda_{580nm}$ = 0.79 nm) was obtained after 120 min indicating no further changes at the device surface, associated with the formation of a dense layer of Affimers. The thickness of this layer was quantified as depicted in Fig. 6.11b (where the contribution of the MPTS layer and the Cu ions was previously substracted), obtaining a value of 3.5 +/− 0.05 nm. This value is in agreement with those values reported in the literature [20]. The potential for creating dense layers of bioreceptors is also certified by the calculated refractive index (Fig. 6.11c), which is 1.445 +/− 0.005 RIU. This value is close to the average refractive index of a dry protein layer such as antibodies ($n_{antibody} \approx$ 1.45 RIU) [18]).

Insight into both the layer formation and kinetics of the process can be obtained from the thickness and refractive index estimations (Figs. 6.11b, and 6.11c, respectively), which helps understading the surface assembly of these bioreceptors. Particularly in this case, I observe three different sub-processes. Firstly, approximately between 30 and 40 min from the beginning of the experiment, a thickness value of 3.45 nm is obtained which corresponds to a refractive index of 1.385 RIU. This relatively low refractive index value is indicative of a low density layer, while in contrast the thickness value is similar to the obtained for a dense layer (as it is at the end of the experiment). Initially, the surface of the sensor is not covered by molecules so that cysteine modified Affimers can easily bind to the Cu^{1+} ions on the surface, without undergoing hydrophobic or electrostatic interactions with adjacent molecules due to the low density of the layer at this stage. In this way, the molecules are placed upright due to their rigidity [19]. Subsequently, between 40 and 74 min from the beginning of the experiment, the thickness of the layer is reduced to 2.3 nm while the refractive index increases to 1.456 RIU. During these instants the refractive index of the layer increases (Δn = 0.071 RIU) as the increment in molecular

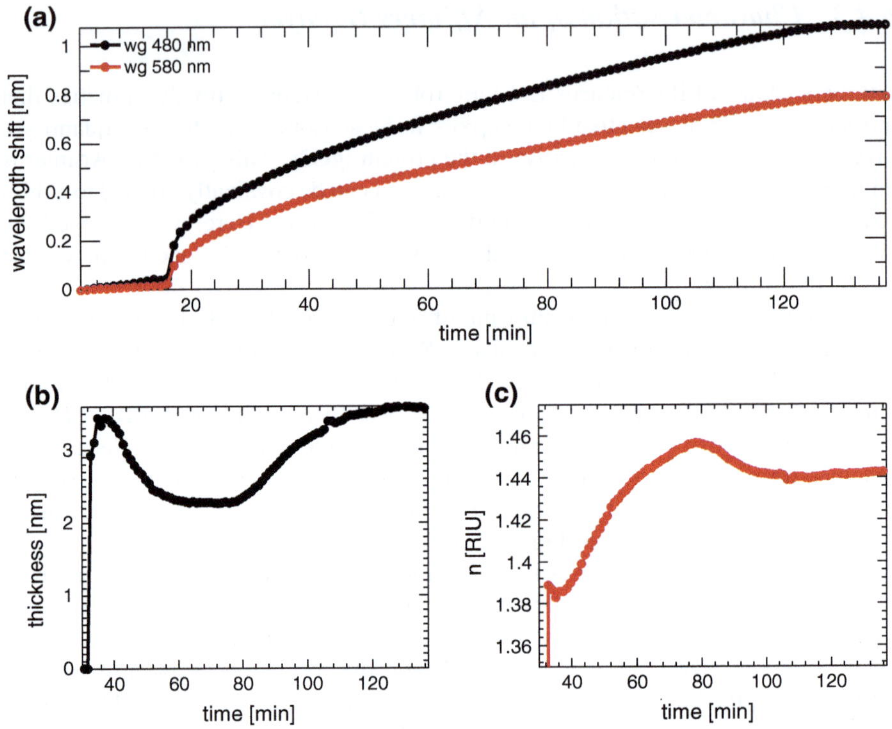

Fig. 6.11 Monitoring of an Affimers layer formation. **a** The resonance wavelength of both 480 and 580 nm width ring sensors was continuously monitored while Affimers bind to the sensor surface. **b** and **c** show the estimations of both thickness and refractive index of the layer, respectively

layer density, hence decreasing the distance between immobilised molecules. This decrement in distance favours hydrophobic and electrostatic interactions between adjacent Affimers molecules [21], therefore the molecules restructure to minimise these interactions and change their orientation (i.e. they are not placed upright). This restructuring results in a decrease of the layer thickness by 1.15 nm. Finally, after 75 min from the beginning of the experiment, the thickness of the layer is gradually increased again to 3.5 nm and the refractive index again decreases to 1.445 RIU. At this point, the density of the layer is further increased until the surface is densely covered with Affimer molecules (due to the strong affinity between the Cu^{1+} ions and the cysteine group, which predominates over hydrophobic and electrostatic interactions). Because of the higher molecular layer density, the Affimer layer restructures once more thus the thickness of the layer increases to 3.5 nm as the molecules are again placed upright.

6.5 Dual-Mode Analysis of Electroactive Hairpin Shaped DNA Strands Conformational Changes

In Sect. 6.4 we have presented the ability of tailoring light-matter interaction to characterise the thicknesses and densities (refractive index) of biolayers. Yet, its potential to simultaneously interrogate in the electrochemical domain in combination with estimation of thicknesses and densities has not been presented. Here we introduce the capability of analysing complex conformational changes of electroactive molecules as is the case of electroactive hairpin shaped DNA strands.

6.5.1 Hairpin Shaped DNA Strands

Hairpin DNA oligonucleotides are DNA strands where two regions of the strand itself undergo base-pair linkage to form a double helix leading to an unpaired loop [22]. These structures are a key building block of many RNA secondary structures and also in the development of DNA computation [23] and DNA machines [24, 25]. They also paved the way for precise sequence-specific detection of DNA through electrochemical interrogation by including an electrochemical probe at the of the sequence [26].

As illustrated in Fig. 6.12a, the structure of DNA harpin strands consists of a loop, a double helix region and, in some occasions, a toehold. The toehold can act as a binding domain that, upon exposure to a complementary invading strand (Fig. 6.12b), may trigger a displacement process which will alter the conformation of the hairpin (Fig. 6.12c) [22]. This process receives the name of toehold-mediated strand displacement, and is often used to drive synthetic nanomachines made from DNA [24, 25].

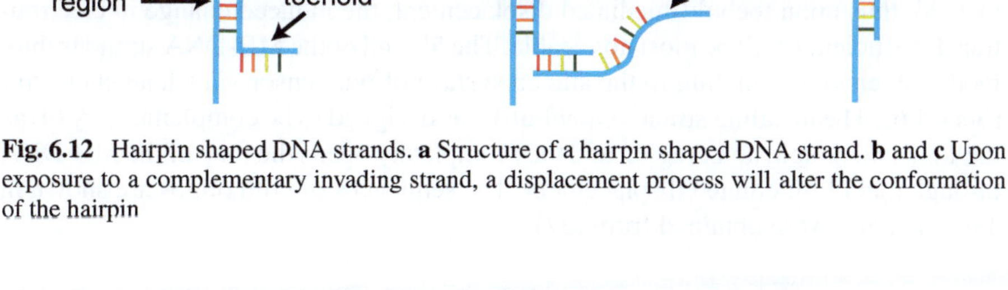

Fig. 6.12 Hairpin shaped DNA strands. **a** Structure of a hairpin shaped DNA strand. **b** and **c** Upon exposure to a complementary invading strand, a displacement process will alter the conformation of the hairpin

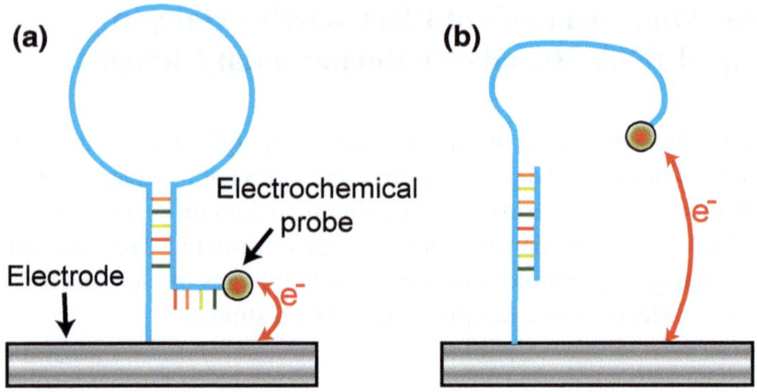

Fig. 6.13 Electrochemical detection of conformational changes in hairpin DNA. **a** Electroactive hairpin DNA structures self-assemble onto an electrode. **b** Upon exposure to a complementary strand, a large conformational change modifies the electron-transfer efficiency

As mentioned previously, harpin DNA oligonucleotides can be modified with an electrochemical probe, allowing electrochemical interrogation by electron tunneling from an electrode into the electrochemical probe. This detection strategy, firstly reported in 2003 [26], involves the immobilisation of electroactive hairpin DNA structures that self-assemble onto an electrode (Fig. 6.13a). Upon exposure to a complementary invading strand, immobilised hairpin DNA strands experience toehold-mediated strand displacement to induce a large conformational change in the immobilised strand (Fig. 6.13b). This conformational change will modify the distance of the electrochemical probe to the electrode, hence changing the electron-transfer efficiency.

6.5.2 Dual-Mode Interrogation

In this section, we particularly focus on investigating, via parallel dual-mode optical and electrochemical interrogation, toehold-mediated strand displacement of a surface-immobilised electroactive DNA hairpin shaped.

The immobilised electroactive hairpin DNA oligo (iba GmbH), MB-DNA, was modified at the 3'- end with the electrochemical probe methylene blue (MB) to allow for electrochemical interrogation. This electroactive end is situated at the end of the toehold, thus upon toehold-mediated displacement, the induced change in electron-transfer efficiency will be most noticeable. The 5'- end of the MB-DNA strand is thiol modified, enabling binding to the silicon surface of our sensor via silane chemistry (Sect. 4.6). The invading strand (OpenFull) was designed to be complementary to the toe-hold and double helix region and thus completely opens the loop of the MB-DNA through toehold-mediated displacement. The sequences of both oligos are shown in Table 6.1, and were obtained from [27].

Table 6.1 Designed DNA oligos sequences for investigating toehold-mediated strand displacement of electroactive hairpin DNA

Label	Sequence
MB-DNA	5'-CTTCTCTATCTTGATGTGAATGAACTGAAC GACGAGCCATCAAGATAGAGAAGGACGAGC-3'
1-2 OpenFull	5'-GCTCGTCCTTCTCTATCTTGATG-3'

The structure of the MB-DNA strand was analysed using NUPACK [28] to validate its conformation before and after exposure to the strand OpenFull. In Fig. 6.14a, the result of the simulation of the MB-DNA oligo structure is shown exhibiting the toehold, the double helix section and the loop. Upon simulated exposure to the invading strand OpenFull, a toehold-mediated conformational change of the MB-DNA is observed in Fig. 6.14b. Neither the toehold nor the loop are observed due to the toehold-mediate displacement. The 3'- end, where the MB molecule is linked, has been distanced from its original location, as required for electrochemical interrogation.

As in Sect. 6.4.3, the silicon surface was firstly thiol-modified by forming a MPTS layer. Secondly, a solution of Cu^{2+} ions was injected to be able to further functionalise the surface with the thiol modified MB-DNA strand. The surface was then washed with PB (pH 7) to remove unreacted Cu^{2+} and a stable resonance wavelength was obtained for both sensors (Fig. 6.15, Step 0). Thereafter, a solution of the MB-DNA (in PB (pH 7)) at a concentration of 1μM was injected at 20 µL/min for 50 min to immobilise onto the sensor surface (Fig. 6.15, Step 1) until a stable wavelength shift was achieved (indicating saturation of molecules on the surface). Finally, 1µM of the invading strand, OpenFull, was flowed for 50 min to allow toehold-mediated displacement and opening of the immobilised hairpin (Fig. 6.15, Step 2). Here again the invading strand was allowed to flow until a stable wavelength shift was achieved. For simplicity, the absolute wavelength shift at the end of each step are shown for this experiment (Fig. 6.15).

In Fig. 6.15, at the end of Step 1, wavelength shifts of $\Delta\lambda_{480nm}$ = 0.28 nm and $\Delta\lambda_{580nm}$ = 0.20 nm are obtained. After resolving the two-system equation relative to this data (accounting for the presence of a combined MPTS and Cu molecules layer), these shifts result in a layer of thickness equal to 1.97 $+/-$ 0.05 nm and a refractive index of 1.427 $+/-$ 0.005 RIU.

The thickness of the layer may also be used to provide insight into the orientation of the immobilised strands. However, it must be noted that a number of factors have to be considered for the following calculations. Firstly, I have assumed that the density of the underlying MPTS and Cu^{+1} ions layer is constant throughout the silicon surface. However, as discussed in Sect. 6.4.2, the structure of the layer is likely to be non-uniform due to the multilayer structure. Secondly, given its relatively low layer density, MB-DNA molecules are not highly packed, thus the lack of electrostatic and hydrophobic interactions results in an inconsistent orientation between MB-DNA molecules.

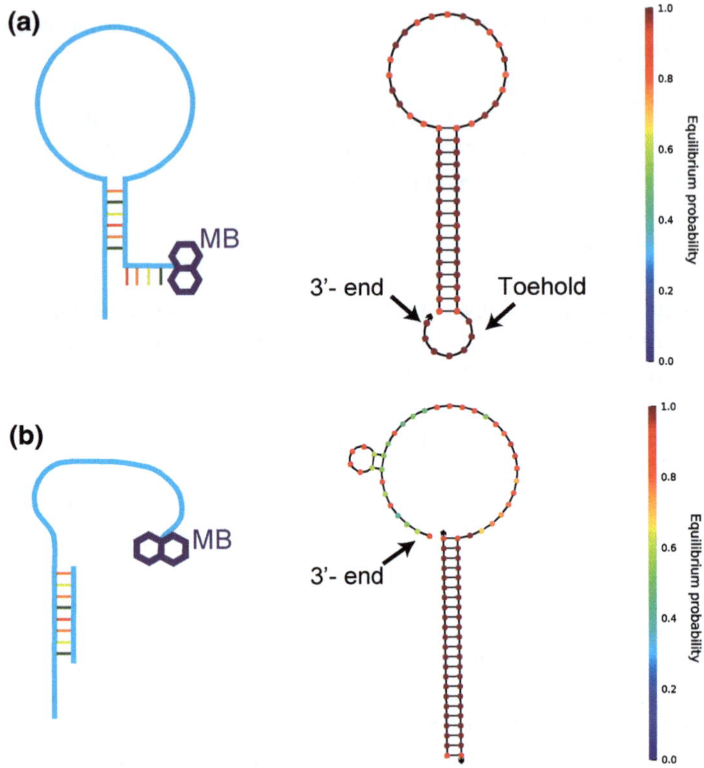

Fig. 6.14 NUPACK analysis of the MB-DNA strand before and after exposure to the strand OpenFull. **a** Before exposure to the strand OpenFull, the MB-DNA oligo structure exhibits a toehold, a double helix section and a loop. **b** Upon exposure to the invading strand OpenFull, a displacement of the 3'- end is observed. Neither the toehold nor the loop are further observed

The hairpin strand MB-DNA, shown in Fig. 6.14a, compromises a 16 base pair (bp) double helix section and a loop 21 bp long. Given the finite length of a base pair, which equals 0.34 nm [29], the length of the strand double helix section of MB-DNA can be approximated to 5.44 nm (Fig. 6.16). Assuming that the single stranded loop section is polymorphically oriented [30], I assume a persistence length of 0 nm for the single stranded region in the following calculations [31]. Therefore, considering the thickness of the layer and the length of the molecule, the angle at which the molecules are oriented respect to the silicon surface is calculated to be $\alpha_{step1} = 21.2°$.

Via electrochemical interrogation, the (bio)chemical activity of the layer can be assessed before and after the toehold-mediated displacement. In this case, through the method detailed in Sect. 4.6, the density of molecules can be directly and unambiguously measured. Figure 6.17 shows the two voltamogramms (at 50 mV/s scanrate, with PB as buffer) related to both Step 1 and Step 2 of the displacement process.

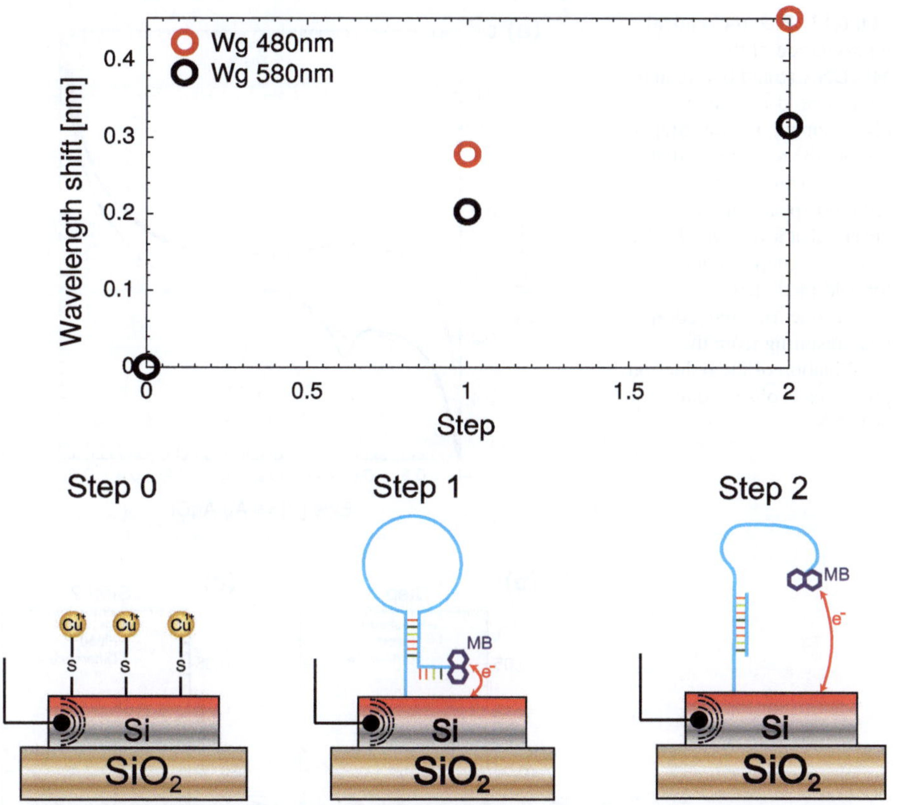

Fig. 6.15 Optical monitoring response analysis of a toehold-mediated conformational change of the MB-DNA. *Step 0* the surface is functionalised with Cu^{+1} ions to enable linkage to thiol group. *Step 1* The MB-DNA strand is immobilised on the sensors. *Step 2* Toehold-mediated displacement triggered by the strand OpenFull results in a conformational change of the MB-DNA strand. Error bars have not been included in this figure as they are smaller than the represented data points

Fig. 6.16 Conformation analysis of the MB-DNA strand before toehold-mediated displacement. Before the toehold-mediated displacement, the MB-DNA oligos are oriented at α_{step1} = 21.2° with the surface

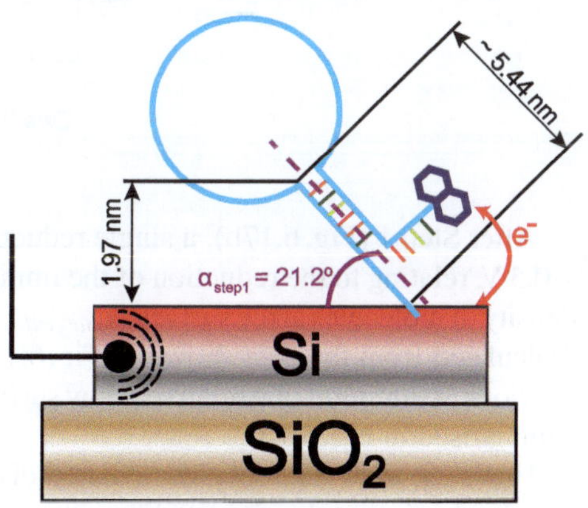

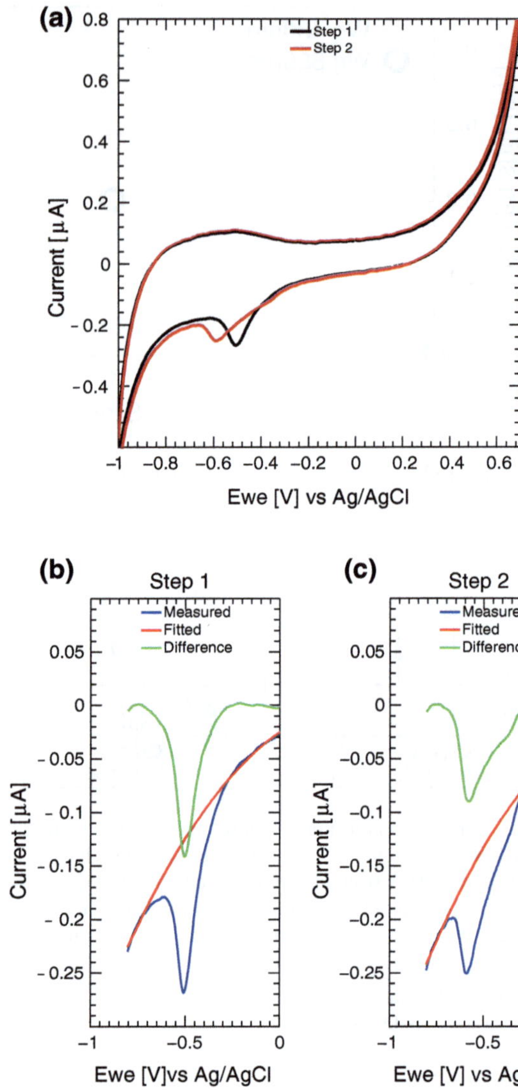

Fig. 6.17 Electrochemical interrogation of the MB-DNA strand before and after toehold-mediated displacement. **a** After Step 1 (before the toehold-mediated displacement), a single reduction peak can be observed at a voltage of −0.5 V. **b** After Step 2 (after the toehold-mediated displacement), a reduction peak resulting from the combination of the reduction processes is observed at −0.55 V

After Step 1 (Fig. 6.17b), a single reduction peak can be observed at a voltage of −0.5 V, relating to the reduction of the immobilised MB molecule. At this point, the density of molecules at the surface, $n_{molecules Step1}$, equals to 1.1×10^{11} molecules/cm^2 (calculated from the data shown in Fig. 6.17b). This value is in agreement with the previous assumption of a relatively low surface density employed for the calculations summarised in Fig. 6.16.

However, after the toehold-mediated displacement (Step 2 in Fig. 6.17c, a combined reduction peak originated from two reduction processes at two different voltages (approximately −0.6 and −0.475 V) is observed. This combination of peaks indicates that not all the immobilised MB-DNA strands have undergone toehold-mediated displacement, therefore undisplaced MB-DNA strands are being reduced at −0.475 V while displaced MB-DNA strand are being reduced at −0.6 V. Here we present again how the combination of the optical and electrical domains results in a powerful tool to obtain deeper insight into complex (bio)molecular processes. The density of molecules at the surface after Step 2, $n_{moleculesStep2}$, equals to 8.9×10^{10} molecules/cm^2, which signifies that immobilised MB-DNA strands after Step 1 have not been released from the surface. I note that the small difference between $n_{moleculesStep1}$ and $n_{moleculesStep2}$ is due to the accuracy of the fitting tool when substracting the background current. At this stage, it is also possible to calculate the fraction of the MB-DNA molecules that have undergone toehold-mediated displacement by calculating the number of molecules in each of the reduction peaks shown in Fig. 6.17c. After fitting two independent gaussians to each reduction process and integrating the area under each curve, we estimate that a fraction of 74.2% of the immobilised MB-DNA strands have been displaced.

Finally, there is a reduction in the amplitude of the reduction peak following the opening of the immobilised MB-DNA hairpin. This reduction in current amplitude, coupled with the shift in reduction voltage ($\Delta V = -0.05$ V), indicates that the tunneling efficiency between the substrate and the MB probe has decreased. MB is extremely sensitive to variations in its close environment (such as variations in the pH level), and therefore conformational changes in the molecule it is linked to will have an impact on its tunneling efficiency (see [32] for details about the use of MB as redox-active indicator for probing molecular interactions). In this case, the alteration in the MB environment was originated from a change in the layer structure after exposure to the strand OpenFull (end of Step 2). However, the electrochemical information obtained to this extend does not provide clear evidence of the new structure of the layer.

Considering the optical data shown in Fig. 6.15, it is possible to obtain insight into the new structure of the MB-DNA layer at this stage of the experiment, which will complement the information provided by the electrochemical domain. From the obtained optical shifts, $\Delta\lambda_{480nm} = 0.46$ nm and $\Delta\lambda_{580nm} = 0.32$ nm, the effect in the MB-DNA layer structure originated after exposure to the strand OpenFull can be identified. These optical shifts result in a layer thickness of 2.21 +/− 0.05 nm and a refractive index of 1.443 +/− 0.005 RIU, which are indicative of changes in both thickness and layer density. The increment in the layer thickness ($\Delta t = 0.24$ nm) suggests that the change in the tunnelling efficiency of the MB-probe was due to a change in position of the MB probe in respect to the electrode/sensor surface. Alternatively, the variation in the refractive index of the layer ($\Delta n = 0.016$ RIU) also denotes a change in the layer structure. As mentioned before, the density of the layer has not increased as the number of reduced MB molecules does not significantly vary ($n_{moleculesStep1} \approx n_{moleculesStep2}$). We consequently deduce that after Step 2, the increment in the refractive index is due to the change in the structure of the MB-DNA and not to a change in layer density. From the simulations show in Fig. 6.14b, it is appreciated that the structure of the MB-DNA after exposure to the strand OpenFull compromises a 22 bp double helix section (rather than the original 16 bp double helix section), therefore there is an increment in the number of double stranded DNA sections on the surface, hence the increase in the refractive index.

6.6 Comparison to Similar Approaches

In Sect. 6.2, two similar approaches for accurately measuring *t* and *n* of immobilised layers using optical technology were presented. DPI, the first of these two, was proven to be useful in the study of both the impact of small biomolecules on the structure of DNA [33] and the research of protein conformational changes [34]. This technique claims a resolution of 0.1 nm in the context of protein measurements [15] and a maximum estimated error of 0.003 RIU on refractive index and 0.025 pm on thickness for measuring dielectric layers of 8 nm height with the enhanced multiple path length DPI [35]. However, this technique does not allow the implementation of a highly multiplexed sensor platform (due to its large footprint) and has limited throughput. In fact, the sales of this instrument were discontinued in 2012.

Silicon photonics approaches, however, have the potential to miniaturise and further integrate dual polarisation sensing while allowing control on the sensor's geometry. They allow highly multiplexed sensing while exhibiting comparable performance to DPI [7]. The approach based on orthogonally polarised TE and TM modes in a ring resonator presents maximum error of 0.02 nm on thickness estimation and a maximum error on refractive index of 0.00062 RIU for measuring protein conformational changes. Our approach however, present a maximum error of 0.05 nm for the thickness estimation and a maximum error on refractive index of 0.005 RIU. The reason for this loss of accuracy is twofold. Firstly, the fact that two independents sensor are used for the *t* and *n* estimations implies that the noise of each sensor is convoluted with the error of the other sensor. And also, differences between overlayers immobilised on each of the sensors will provide wrong estimations. Secondly, the difference in sensitivity (nm/nm, Fig. 6.2) between both mode profiles is not as large as the difference in sensitivity between the TE and TM modes [7]. Nevertheless, it should be stated that the actual resolution obtained with our approach fullfils the resolution and error requirements for most of the conformational analysis reported in the literature using either DPI or orthogonally polarised TE and TM modes in a ring resonator [6, 7, 15, 34]. Finally, this approach allows for simultaneously interrogating the electrochemical properties of the bound layer of molecules, which, to our best knowledge, has not been performed before.

6.7 Conclusions

The ability of our platform for tailoring the light-matter interaction is described and used to quantify the refractive index and thickness of immobilised biolayers. We proposed a cascaded dual ring resonators approach in which the distribution of the optical field is different for each of them, hence behaving differently when biolayers are immobilised on them. We used this difference in behaviour of the rings to formulate a two-equation system which can be uniquely resolved for an specific thickness and refractive index. Biolayers of MPTS and Affimers have been characterised, while the ability to measure complex structural conformational changes in electroactive hairpin shaped DNA strands has been demonstrated combining electrochemical and optical information. Silicon electro-photonic sensors thus offer a highly multiplexed platform to study complex biomolecular interactions in more depth than is possible with single-mode sensors.

References

1. X. Fan, Y. Sun, Optical ring resonators for biochemical and chemical sensing. Anal. Bioanal. Chem. **399**, 205–211 (2011)
2. J. Homola, S. Yee, G. Gauglitz, Surface plasmon resonance sensors: review. Sens. Actuators **54**(54), 3–15 (1999)
3. Yuanjin Zhao, Gu Xiangwei Zhao, Zhongze, Photonic crystals in bioassays. Adv. Funct. Mater. **20**(18), 2970–2988 (2010)
4. R.G. Heideman, P.V. Lambeck, Remote opto-chemical sensing with extreme sensitivity: design, fabrication and performance of a pigtailed integrated optical phase-modulated machzehnder interferometer system. Sens. Actuators B Chem. 61 (13),100–127 (1999), http://www.sciencedirect.com/science/article/pii/S092540059900283X
5. S.E. Radford, Protein folding: progress made and promises ahead. Trends Biochem. Sci. **25**(12), 611–618 (2000)
6. N. J. Freeman, G. H. Cross, M. J. Swann. *Dual Polarization Interferometry: A Real-Time Optical Technique for Measuring (Bio)molecular Orientation, Structure and Function at the Solid/Liquid Interface*. Wiley (2008)
7. S. Werquin, J.W. Hoste, P. Bienstman, Conformational analysis of proteins with a dual polarisation silicon microring. Opt. Express **22**(3), 2807–2820 (2014)
8. A. N. Parikha, D. L. Allarac, *Handbook of Biofunctional Surfaces*. Pan Stanford Publishing (2013)
9. K. Okamoto, *Fundamentals of Optical Waveguides*. Elsevier (2006)
10. W. Bogaerts, P. De Heyn, T. Van Vaerenbergh, K. De Vos, S. Kumar, Silicon microring resonators. Laser Photonics Rev. **6**(1), 47–73 (2012)
11. K. De Vos, Label-free silicon photonics biosensor platform with microring resonators. PhD thesis, University of Ghent (2010)
12. S. Johnson, D. Evans, S. Laurenson, D. Paul, A.G. Davies, P. Ko Ferrigno, C. Wlti, Surface-immobilized peptide aptamers as probe molecules for protein detection. Anal. Chem. **80**, 978–983 (2008)
13. S.H. Brorson, Bovine serum albumin (BSA) as a reagent against non-specific immunogold labeling on lrwhite and epoxy resin. Micron **28**, 189–195 (1997)
14. D.C. Carter, J.X. Ho, Structure of serum albumin. Adv. Protein Chem. **45**, 153–203 (1994)
15. M.J. Swann, G.H. Cross, A. Reeves, S. Brand, N.J. Freeman, L.L. Peel, J.R. Lu, Real time, high resolution studies of protein adsorption and structure at the solidliquid interface using dual polarization interferometry. J. Phys. Condens. Matter **16**(26), S2493 (2004)
16. J-W. Hoste, Dual Polarization Ring Resonator for Conformational Analysis of Molecular-sized Layers. PhD thesis, Universiteit Gent (2015)
17. Vincent S. Stoll, John S. Blanchard. [4] buffers: Principles and practice. In: Murray P. Deutscher (ed.), Guide to Protein Purification, vol. 182 of Methods in Enzymology, pp. 24–38. Academic Press (1990)

18. A. Ulman, Formation and structure of self-assembled monolayers. Chem. Rev. **96**, 1533–1554 (1996)
19. R. Woodman, M. Crawford, P. K. Ferrigno, Peptide aptamers: Tools for biology and drug discovery. 2(1), 72–79 (2003)
20. X. Luo, J. Davis, Electrical biosensors and the label free detection of protein disease biomarkers. Chem. Soc. Rev., 5944–5962 (2013)
21. D. Chandler, Interfaces and the driving force of hydrophobic assembly. Nature **437**, 640–647 (2005)
22. R.G. Berendsen, H.G.J.M. Janssen, T.J.J. Binnendijk, C.W. Hilbers, G.A. van der Marel, C.A.G. Haasnoot, kinetics and thermodynamics of dna hairpin fragments in solution. J. Biomol. Struct. Dyn. **1**(1), 115–129 (1983)
23. J. Bath, A. Tyrrell, M. Costa, A clocked finite state machine built from DNA. Chem. Commun. **49**, 237–239 (2013)
24. R. Sha, Omabegho, To. N. C. Seeman. A bipedal dna brownian motor with coordinated legs. Science **324**(5923), 67–71 (2009). 10.1126/science.1170336
25. J. Bath, S. J. Green, A. J. Turberfield, Coordinated chemomechanical cycles: A mechanism for autonomous molecular motion. Phys. Rev. Lett. **101**, 238101 (2008)
26. K. W. Plaxco, C. Fan, A. J. Heeger. Electrochemical interrogation of conformational changes as a reagentless method for the sequence-specific detection of DNA. Proc. Nat. Acad. Sci. **100**(16), 9134–9137 (2003)
27. D. Lubrich, S.J. Green, A.J. Turberfield, DNA hairpins: Fuel for autonomous DNA devices. Biophys. J. **91**(8), 2966–2975 (2006)
28. J.S. Bois, B.R. Wolfe, M.B. Pierce, A.R. Khan, R.M. Dirks, J.N. Zadeh, C.D. Steenberg, N.A. Pierce, Nupack: Analysis and design of nucleic acid systems. J. Comput. Chem. **32**(1), 170–173 (2011)
29. W. N. Hunter, G. Kneale, T. Brown, O. Kennard. Molecular structure of the G.A base pair in DNA and its implications for the mechanism of transversion mutations. Proc. Natl. Acad. Sci. USA, 83(8), 2402–2406 (1986)
30. A.C. Gustafsson, F. Sterky, P. Nyrn, M. Uhln, A. Ahmadian, Baback Gharizadeh and J. Lundeberg. Single-nucleotide polymorphism analysis by pyrosequencing. Anal. Biochem. **280**(1), 103–110 (2000)
31. G. Wang, Q. Chi, J. Jiang, The persistence length and length per base of single-stranded DNA obtained from fluorescence correlation spectroscopy measurements using mean field theory. Physica A **392**(5), 1072–1079 (2013)
32. D. Langford, R. Bonand, E. Koutsoumpeli, J. Murray, S. Johnson, Probing molecular interactions with methylene blue derivatized self-assembled monolayers. Sens. BioSens. Res., pp. 1–6 (2015)
33. Z. Zhang F. Yang J. Wang, X. Xu and Xi. Yang. Real-time study of genomic dna structural changes upon interaction with small molecules using dual-polarization interferometry. Anal. Chem. 81(12), 4914–4921 (2009)
34. T.A. Jowitt, J. Ylstalo, P. Coffey, R.S. Meadows, L. Ala-Kokko, D.J. Thornton, M. Fresquet, M.D. Briggs, Structural and functional characterization of recombinant matrilin-3 a-domain and implications for human genetic bone diseases. J. Biol. Chem. **282**(48), 34634–34643 (2007)
35. T.A. Waigh, F. Schedin, P.D. Coffey, M.J. Swann, J.R. Lu, Multiple path length dual polarization interferometry. Opt. Express **17**(13), 10959–10969 (2009)

Chapter 7
Conclusions and Outlook

This chapter provides a coherent summary of the research presented in this thesis. In addition, the prospects and further applications of the sensor developed in this thesis are discussed.

7.1 Summary and Conclusions

Several single-domain techniques already exist which provide accurate, fast and reliable analysis of chemical events at the surface of a sensor (Chap. 2). Nevertheless, techniques that monitor such chemical or biological events in two measurement domains (such as EC-SPR, E-QCM-D and EC-OWLS) are becoming increasingly popular within the scientific community as they provide further insight into the underlying molecular and biomolecular processes. Here, I have demonstrated a technology that integrates silicon photonics with electrochemistry into a single, innovative technology (Chap. 4). The novelty of this technology is that it provides additional functionality that until now been inaccessible to silicon photonics (Chaps. 5 and 6).

To allow the integration of electrochemical and photonic sensing techniques on the same platform, an approach whereby the properties of an optical substrate (here SOI) have been controlled such that electron transfer is possible. By doping the silicon substrate, it is possible to tune its conductivity sufficiently to enable its use as an electrochemical electrode. A conflict exits in this approach as the electrical and optical properties of the silicon substrate are related. For example, a highly doped substrate optimised for high electrical conductivity has a negative impact on the optical properties. This issue is attributed to the doping process which increases the number of impurities in the crystalline silicon layer, in turn increasing the free carrier loss [1, 2]. This increase impacts directly on the performance of a resonant cavity. A solution based on the control of the spatial distribution of dopants in the silicon device layer is presented. Following this solution, a highly doped film

J. Juan Colaś , *Dual-Mode Electro-photonic Silicon Biosensors*, Springer Theses,
DOI 10.1007/978-3-319-60501-2_7

(N_D =1 × 10^{18-20}cm^{-3}) is obtained in the top 10 nm of the silicon layer, sufficient to support electron transfer to/from a solution-phase analyte. From an optical viewpoint, this minimises the interaction of the mode with the doped region. The overall optical loss is thus minimised. The final device, based on a photonic microring resonator presents an average doping density of N_D =7.5 × 10^{16}cm^{-3} and a Q-factor of the order of 50,000. An assessment of the correlation between the doping and the Q-factor was also provided, which models our system extremely well. The sensitivity of the device, measured to be 65.1 nm/RIU, leads to a *LOD* 1.15 × 10^{-5}, comparable to ring resonator sensors fabricated in industrial foundries and published in the literature. The electro-photonic sensor discussed here presents similar sensitivity levels to E-SPR and E-QCM-D systems [3]. A proof-of-concept experiment was carried out to monitor and quantify, from both electrochemical and optical domains, the immobilisation of a layer of redox active molecules (here a MB-SAM) on the surface of our sensor. We calculated a molecular layer density of 2.2 × 10^{12}cm^2 which compares well with the surface density of the same MB-SAM assembled on gold. This precise quantitative information about the monolayer is not possible with photonic detection alone unless ambiguous assumptions are made. We note, in the current proof-of-concept device, continuous application of high voltage values (above 2 V) induce an optical shift that will interfere with the optical measurement signal. Furthermore, the approach is limited to the study of surface-immobilised (bio)molecules.

Our dual-mode sensor not only combines electrochemical characterisation with photonic sensing, but also enables electrochemical processes for selective modification of the silicon surface to regulate the local surface chemistry (Chap. 5). We showed how electrografting of diazonium ions provides well-ordered and homogeneous monolayers of controlled structure for functionalising the surface of our sensor. Three different electrografted layers of in situ generated diazonium ions were studied, namely 4-ethynylbenzene diazonium, n-(4-aminophenyl)maleimide and 4-azidoaniline. We showed how the combination of photonics and electrochemistry provides additional insight into these reactions, as we were able to distinguish between monolayer and multilayer assemblies. Monolayers of electrografted diazoniums of 4-ethynylbenzene diazonium and n-(4-aminophenyl)maleimide were studied, providing alkyne and maleimide biomodifications to the surface, respectively, with molecular layer densities of 3.7 × 10^{13} and 6.2 × 10^{13} molecules/cm^2, respectively. In contrast, the electrografting of 4-azidoaniline resulted in a multilayer structure. Functionalisation of electrografted layers of diazonium ions was demonstrated for the n-(4-aminophenyl)maleimide and 4-azidoaniline diazonium, using the thiol-maleimide and the CuAAC click reaction, respectively. Due to the fact that electrografting is underpinned by electron tunnelling, the reaction is highly spatially localised and occurs only at the electrode. This can be exploited to enable surface modification with high spatial resolution. This reaction was used to develop a proof-of-concept multiplexed photonic sensor array composed of two identical ring resonators, with which we have been able to selectively control (temporally and spatially) the immobilisation of receptor molecules on each of our sensors at a micron scale. The sensors were selectively functionalised via electroreduction of the diazonium ions 4-ethynylbenzene and 4-azidoaniline, and through the CuAAC click reaction, two single stranded DNA molecules of different sequences were selectively immobilised in each of the sensors. Subsequently we demonstrated multiplexed detection of DNA-bioconjugation.

Most evanescent wave optical sensors are sensitive to the combined change in the refractive index and thickness of a biolayer. It is thus not possible to distinguish between these two effects. The platform presented here allows tailoring and optimisation of the light-matter interaction, something that cannot be done using other combined electrochemical-optical sensors such as EC-SPR sensors. We used this capability for the development of a new configuration of photonic ring resonator sensors that enables the quantification of both thickness and refractive index of a biolayer of molecules (Chap. 6). In this configuration, two cascaded ring resonators with waveguide widths of 480 and 580 nm, respectively, respond differently to the same biolayer due to differences in the optical field distribution. The effect of changes in either the refractive index or the thickness of a biolayer on a ring resonator is modelled for the two different waveguide profiles. Since each waveguide profile will provide a different response to a given biolayer, it is possible to disentangle the refractive index and thickness of an overlayer via a two-equation system, and therefore yield information on the molecular structure of biolayers. The assembly of a MPTS SAM was characterised resulting in a thickness of 4.32 ± 0.05 nm and a refractive index of 1.431 ± 0.01 RIU, which indicates that a multilayer structure has been built. A monolayer of affimer bioreceptors was also characterised, obtaining a layer thickness of 3.5 ± 0.05 nm and a refractive index of 1.465 ± 0.01 RIU, indicative of a dense monolayer assembly. Due to its accuracy and capability for parallel bi-domain interrogation, our device has also the prospect of being able to provide insight into complex conformational changes. We proved how our technology is well positioned for the study of conformational changes in hairpin shape DNA strands. The effect of toehold-mediated displacement in an electroactive hairpin shape DNA strand was investigated, revealing the change in thickness of the strand before and after displacement.

7.2 Outlook

Finally, I believe that while this thesis introduces a novel diagnostic technology for label-free sensing, and has presented examples that indicate its weaknesses and strengths, further control of the sensor environment (e.g. temperature) needs to be incorporated to increase stability and reproducibility within experiments. Provided that our technology is fabricated entirely using CMOS compatible materials and processes, it is conveniently situated for the fabrication of low-cost point-of-care devices. I would also consider different approaches for incorporating microfluidics to this platform to allow to device to be recycled, to simplify use of the device and to enable the introduction of clinical samples.

The ability to direct the immobilisation of different biomolecules onto individual sensors within a silicon photonic array at an order of magnitude lower than other methods for the fabrication of photonic arrays (e.g. microfluidic masking, ink-jet printing), can underpin the fabrication of very high-density photonic sensor arrays. In this way one can envisage a sensor with a total area of only a few square micrometers, yet with multiple different sensing sites, such that multiplexed sensing of panels of biomarkers or multimodal sensing inside eukaryotic cells becomes a realistic possibility. Furthermore, each of the different sensing sites can be seen as a different electrode, which can be used for further interrogating the electrochemical properties of the bound molecules via impedance spectroscopy.

The ability to monitor electrochemical processes in situ with photonic sensing also broadens the range of systems that are accessible for silicon photonics. For example, for the first time, our technology will permit simultaneous measurements of enzyme activity (via the electrochemical domain) in parallel with quantification of substrate binding (photonic domain).

This technique could be also be implemented using a 1D photonic crystal resonant grating as sensor, which would allow for spatially imaging the secretion of different proteins from a living cell. Due to the high spatial and temporal resolution of this sensor, specific localised biochemical events can be monitored in real-time. In this manner, different areas within the grating (micrometers away from each other) could be functionalised with a different bioreceptor, hence being selective to specifically secreted proteins.

References

1. R.N. Thomas, D.K. Schroder, J.C. Swartz, Free carrier absorption in silicon. IEEE J. Solid-State Circuits **13**(1), 180–187 (1978)
2. J. Isenberg, W. Warta, Free carrier absorption in heavily doped silicon layers. Appl. Phys. Lett. **84**(13), 2265–2267 (2004)
3. W. Bogaerts, P. De Heyn, T. Van Vaerenbergh, K. De Vos, S. Kumar, T. Selvaraja, P. Claes, P. Dumon, D.Van Bienstman, Thourhout, and R. Baets, Silicon microring resonators. Laser Photonics Rev. **6**(1), 47–73 (2012)

Curriculum Vitae

Dr. José Juan-Colás

Multidisciplinary researcher interested in developing biosensing technologies that impact across health science, clinical practice and chemistry.

Education

2013–2016 **PhD in Electronic Engineering, University of York, York (UK)** *Thesis title: "Dual-mode electro-photonic silicon biosensors"* Supervisors: Dr. Steven D. Johnson and Prof. Thomas F. Krauss

2010–2013 **Telecommunications Engineer, Universitat Politècnica de València, València (Spain)** Final degree project: "Ring resonator (bio)sensor with integrated multichannel spectrometer read-out" Supervisor: Prof. Wim Bogaerts

2007–2010 Technical Telecommunications Engineer. Specialization in Telecommunications Systems., Universitat Politècnica de València, Gandia (Spain) Final degree project: "Design and implementation of control and synchronization processes in a quantum key distribution system in fiber optics based on subcarrier multiplexing" Supervisor: Dr. Alfonso Martínez-Garcia

Experience

2016–Date **Post-doctoral Research Associate** *Mapping chemokine gradients in real-time using novel nano-photonic and electronic devices* MRC Discovery Grant MC PC 15073Departments of Physics and Electronics, University of York (UK)

J. Juan Colaś, *Dual-Mode Electro-photonic Silicon Biosensors*, Springer Theses,
DOI 10.1007/978-3-319-60501-2

2015–2016 **Laboratory demonstrator** *Nanofabrication course* Department of Electronics, University of York (UK)

2012–2013 **Erasmus placement for the elaboration of the master thesis** PHOTONICS RESEARCH GROUP (INTEC) University of Gent (Belgium) Title: "Ring resonator (bio) sensor with integrated multichannel spectrometer read-out". Completely developed and tested an optical device configuration which boosts the sensitivity of ring resonator (bio)sensors.

2012 **Collaboration grant with the Photonics research group** DEPARTMENT OF COMMUNICATIONS Universitat Politècnica de València (Spain) Supervisor: Prof. Daniel Pastor "Becas de colaboración 2011" (EDU/1868/2011), Spanish's Ministry of Education, Culture and Sport. Developed a 2D OCDMA PON Monitoring system simulator.

2010 **Conduct of an internship in a company** GABINETE TÉCNICO PEPSAL, S.L, AND RESEARCH INSTITUTE ITEAM València-Murcia (Spain) Collaboration agreement (code AS14706) established between the Universitat Politècnica de València and the company (800 hours performed). Developed and tested a SCM-QKD control system for future information technology based on quantum optical systems.

Conferences and Publications

2016 **'The electrophotonic silicon biosensor' J. Juan-Colás**, A. Parkin, K. E. Dunn, M. G. Scullion, T. F. Krauss and S. D. Johnson, *Nature Communications* **7:12769** doi:10.1038/ncomms12769

2016 **Biosensors 2016** INVITED TALK *Multi-domain, electro-photonic biosensors* Gothenburg (Sweden)

2016 **Europtrode XII** ORAL PRESENTATION *Multi-domain, electro-photonic biosensors* Graz (Austria)

2015 **"Microring Resonator Refractive Index Sensor with Integrated Spectrometer"** A. Ruocco, J. Juan Colás, W. Bogaerts
12th International Conference on Group IV Photonics (GFP), Canada.

2010 **"Demostración experimental de la viabilidad de Sistemas Fotónicos de Distribución Cuántica de Clave con Multiplexación de Subportadora"** A. Ruiz-Alba Gayá, J. Mora Almerich, V. García Munoz, W. Amaya Ocampo, J. Juan Colás, J. Capmany Francoy
XXV Simposium Nacional Unión Científica internacional de Radio (URSI 2010), Bilbao (Spain).

Awards

2017 **Springer Theses recognition for outstanding Ph.D. research** Springer International Publishing AG. Part of Springer Nature

2016 **Nature Photonics prize for the best poster in a conference** Photonic and Electromagnetic Crystal Structures (PECS) XII, York

2015 **Second prize in the departmental poster competition** The Department of Electronics, York

2013 **Award for outstanding academic performance (state level)** Generalitat Valenciana, València

2013 **Award for the best academic and human performance** Consejo Social de la Universitat Politècnica de València, València

2011 **Award for the best graduate of the Polytechnic School of Gandía** Universitat Politècnica de València, Gandía

2010 **Award for the best final degree project and academic transcript of Spain** Colegio Oficial de Ingenieros Técnicos de Telecomunicación, Madrid

2010 **Award for the best academic and human performance** Consejo Social de la Universitat Politècnica de València, Gandía

2010 **Award for the best final degree performed in a company** Bancaja/Universitat Politècnica de València, València

2010 **Award for the best graduate in Technical Telecommunications Engineer of the Polytechnic School of Gandía. Specialization in Telecommunications Systems** Universitat Politècnica de València, Gandía

2010 **Award for the student with the best academic transcript of third course** Universitat Politècnica de València, Gandía

2009 **Award for the student with the best academic transcript of second course** Universitat Politècnica de València, Gandía

2009 **Award for the student with the best academic transcript of first course** Universitat Politècnica de València, Gandía

MIX
Papier aus verantwortungsvollen Quellen
Paper from responsible sources
FSC® C105338

If you have any concerns about our products,
you can contact us on
ProductSafety@springernature.com

In case Publisher is established outside the EU,
the EU authorized representative is:
Springer Nature Customer Service Center GmbH
Europaplatz 3, 69115 Heidelberg, Germany

Printed by Libri Plureos GmbH
in Hamburg, Germany